TABLE OF CONTENTS

Understanding Levitation

Levitation, the ability to defy gravity and float in mid-air, has captivated human imagination for centuries. From ancient myths and legends to modern-day magic shows, levitation continues to intrigue and inspire. In this article, we will delve into the concept of levitation, exploring its history, types, and the benefits associated with this extraordinary phenomenon.

History of Levitation: The concept of levitation dates back to ancient times, with references found in various cultures and religious texts. In Hinduism, for example, yogis and sages were said to possess the power of levitation, demonstrating their advanced spiritual abilities. Similar accounts can be found in ancient Chinese and Egyptian texts, as well as in the folklore of other civilizations.

Types of Levitation: Levitation can be categorized into different types based on the entities involved. One form is object levitation, where individuals demonstrate the ability to lift and suspend objects in mid-air without any apparent physical support. Another type is self-levitation, where individuals are able to float or hover above the ground or other surfaces. Both forms of levitation have been reported throughout history, often associated with mystical or supernatural powers.

Benefits of Levitation: Beyond its mesmerizing visual appeal, levitation is believed to offer several benefits. From a spiritual perspective, levitation can be seen as a manifestation of deep meditation and a heightened connection with higher realms of consciousness.

Practitioners often describe a sense of weightlessness and a feeling of expansion and liberation during levitation experiences. Additionally, levitation can serve as a powerful tool for personal growth, helping individuals develop focus, discipline, and a greater understanding of energy manipulation.

Preparing for Levitation: Achieving levitation requires a combination of physical, mental, and environmental preparation. Physically, it is essential to maintain a good level of fitness and flexibility, as well as strengthen the core muscles to support the body during levitation. Mentally, cultivating a calm and focused mind through meditation and visualization techniques can enhance the ability to achieve levitation. Creating a serene and harmonious environment, free from distractions, is also crucial for successful levitation practices.

Developing Levitation Techniques: To master levitation, it is necessary to learn and practice specific techniques. Grounding and centering exercises help individuals establish a strong connection with the Earth's energy, providing a stable foundation for levitation. Meditation and visualization techniques aid in developing concentration and the ability to envision oneself floating effortlessly. Energy manipulation exercises enable individuals to harness and direct their internal energy, facilitating levitation.

Levitation Exercises for Beginners: For those new to levitation, there are several exercises to start with. The floating feather technique involves focusing on a lightweight object, such as a feather, and visualizing it floating gently in the air. The balloon lift exercise requires individuals to imagine themselves being lifted by balloons, gradually increasing the sensation of weightlessness. The levitation cushion practice involves sitting on a cushion and visualizing the cushion supporting and lifting the body.

Advanced Levitation Techniques: Once basic levitation skills are mastered, individuals can progress to more advanced techniques. These may include levitating objects, such as small items or even larger furniture, through focused intention and energy manipulation. Self-levitation techniques involve suspending oneself in mid-air through a combination of mental focus, energy control, and physical alignment. Advanced practitioners may even explore levitation while in motion, moving gracefully and effortlessly through the air.

In conclusion, levitation is a captivating phenomenon that has fascinated humanity for ages. While it may seem otherworldly, with dedication and practice, individuals can unlock the potential to levitate. Beyond the visual spectacle, levitation offers a path of spiritual growth, expanded awareness, and personal development. By understanding the history, types, and benefits of levitation, one can embark on a transformative journey towards mastering this extraordinary ability.

History of Levitation

Levitation, the seemingly magical act of defying gravity and floating in mid-air, has a long and fascinating history that spans across different cultures and civilizations. From ancient legends to modern-day performances, the concept of levitation has captivated the human imagination for centuries. In this article, we will explore the intriguing history of levitation, tracing its origins and evolution through time.

Ancient Beginnings: The origins of levitation can be traced back to ancient civilizations where it was often associated with mysticism, spirituality, and supernatural powers. In ancient Hindu texts, known as the Vedas, there are references to yogis and sages who possessed the ability to levitate. These individuals were believed to have attained advanced levels of spiritual enlightenment and were revered for their mystical powers.

Similar accounts of levitation can be found in ancient Chinese texts, particularly in Taoist and Buddhist traditions. Taoist practitioners sought to cultivate a harmonious balance between nature and the self, and levitation was seen as a manifestation of spiritual attainment and mastery over the physical realm. In Buddhism, there are legends of monks who achieved levitation through deep meditation and mindfulness practices.

Medieval Europe and Levitation: During the Middle Ages, levitation became associated with religious figures and saints. Numerous accounts and hagiographies described saints and ascetics levitating during moments of intense prayer or divine ecstasy. These accounts served to reinforce the belief in the supernatural and the power of faith.

One of the most well-known figures associated with levitation is St. Joseph of Cupertino, an Italian friar from the 17th century. He was said to have experienced frequent levitation episodes during prayer and religious ceremonies. His levitations were witnessed by many, including high-ranking church officials and royalty, leading to his eventual canonization as a saint.

Levitation in Eastern Practices: In Eastern cultures, particularly in India and Tibet, levitation was intertwined with the practices of yoga and meditation. Yogis and sadhus were believed to possess the ability to levitate through intense concentration, control of breath, and mastery over their physical and energetic bodies. These extraordinary feats of levitation were seen as a demonstration of spiritual prowess and the union of mind, body, and spirit.

In the Tibetan Buddhist tradition, the practice of "trul khor" or "yantra yoga" incorporates physical movements, breath control, and visualization techniques. Through these practices, advanced yogis were said to achieve levitation, demonstrating the profound connection between the mind, energy, and the physical body.

Modern Era and Entertainment: With the advent of stage magic and illusionism, levitation found its place in the world of entertainment. Illusionists and magicians began incorporating levitation acts into their performances, creating a sense of wonder and amazement among audiences. These acts often involve clever tricks, hidden supports, and optical illusions, giving the illusion of floating in mid-air.

In recent times, the art of levitation has continued to evolve with advancements in technology and special effects. Levitation acts have become more elaborate and visually stunning, captivating audiences in theaters, television shows, and even street performances.

In Conclusion: The history of levitation is a tapestry woven with threads of spirituality, mysticism, and human curiosity. From ancient sages and yogis to modern-day illusionists, levitation has fascinated and inspired people across cultures and generations. Whether viewed as a manifestation of divine power, a result of advanced meditation, or a captivating stage performance, levitation continues to ignite our imagination and challenge our understanding of the physical laws that govern our world.

Types of Levitation

Levitation, the remarkable ability to defy gravity and float in mid-air, comes in various forms. Throughout history, different types of levitation have been reported and explored, each with its unique characteristics and manifestations. In this article, we will delve into the fascinating world of levitation and examine some of its distinct types.

1.   Object Levitation: Object levitation refers to the ability to lift and suspend objects in mid-air without any apparent physical support. This type of levitation has been a staple of magic performances and illusionism. Magicians employ a variety of techniques, such as invisible threads, magnets, or mechanical contraptions, to create the illusion of objects floating or moving autonomously. Object levitation is primarily used for entertainment purposes and captivating audiences with seemingly impossible feats.

2.   Self-Levitation: Self-levitation, also known as personal levitation, involves an individual floating or hovering above the ground or other surfaces. This type of levitation has been described in spiritual and mystical contexts, often associated with advanced meditation practices or states of deep trance. Accounts of self-levitation can be found in different cultures and belief systems, including Hinduism, Buddhism, and Taoism. Achieving self-levitation is believed to require exceptional focus, control over energy, and a deep connection with one's inner self.

3.   Magnetic Levitation: Magnetic levitation, also referred to as maglev, is a scientific and engineering phenomenon that utilizes magnetic forces to suspend objects in mid-air. This technology is commonly applied to transportation systems, such as high-speed trains, where the train is levitated and propelled forward using magnetic repulsion and attraction. Maglev technology eliminates friction and allows for smooth, efficient, and high-speed travel. Magnetic levitation is a practical application of levitation principles in the realm of physics and engineering.

4.   Acoustic Levitation: Acoustic levitation is an intriguing form of levitation that employs sound waves to suspend objects in mid-air. By using carefully controlled ultrasonic sound waves, objects can be lifted and held in a stable position against the force of gravity. This technology has various applications, including scientific research, material handling, and even the manipulation of liquid droplets for chemical and biological experiments. Acoustic levitation showcases the ability of sound to counteract gravity and hold objects aloft.

5.   Spiritual Levitation: Spiritual levitation is often associated with mysticism, religious experiences, and heightened states of consciousness. This form of levitation is described in spiritual traditions and mystical accounts as a transcendent phenomenon, where individuals are lifted or transported to higher realms or dimensions. Spiritual levitation is believed to occur through the expansion of consciousness, deep meditation, and the awakening of inner spiritual energies. It is seen as a manifestation of spiritual elevation and liberation from the limitations of the physical body.

6.   Illusionary Levitation: Illusionary levitation, as the name suggests, is a type of levitation that is purely an illusion or trick. This form of levitation is commonly seen in performances, magic shows, and entertainment acts. Illusionists employ various techniques, such as hidden supports, optical illusions, or clever misdirection, to create the appearance of levitation. Illusionary levitation relies on the skillful manipulation of perception and the suspension of disbelief to create awe and wonder among audiences.

In conclusion, levitation encompasses a diverse range of types and manifestations. From object levitation used in magic performances to self-levitation associated with spiritual practices, each type offers a unique perspective on defying gravity. Whether it is through advanced technology, spiritual practices, or the art of illusion, levitation continues to captivate our imagination and challenge our understanding of what is possible in the realm of gravity-defying feats.

Benefits of Levitation

Levitation, the extraordinary ability to float and defy gravity, has fascinated and intrigued people for centuries. While levitation is often associated with magic shows and illusions, it is believed to offer a range of benefits beyond mere entertainment. In this article, we will explore some of the potential benefits associated with levitation, from physical and mental well-being to spiritual growth and self-discovery.

1.   Physical Fitness and Conditioning: Engaging in levitation practices can provide a unique form of physical exercise. The process of achieving levitation requires strength, balance, and control over the body's muscles. Regular practice can help improve core strength, enhance overall body coordination, and increase flexibility. These physical benefits contribute to better posture, improved motor skills, and a general sense of physical well-being.

2.   Mind-Body Connection: Levitation practices often involve a deep focus on the mind-body connection. By cultivating awareness and concentration, individuals learn to align their mental intention with physical action. This heightened mind-body connection can lead to increased mindfulness, improved concentration, and a greater sense of control over one's physical and mental states. The discipline and focus required for levitation can extend beyond the practice itself, positively influencing other areas of life.

3.   Expanded Consciousness: Levitation experiences have been associated with expanded states of consciousness and altered perceptions. During levitation, individuals often report a heightened sense of awareness, a feeling of liberation from the constraints of the physical body, and a connection to higher realms of consciousness. These experiences can lead to profound spiritual insights, a deepening of one's understanding of oneself and the universe, and a greater sense of purpose and interconnectedness.

4.   Energy Manipulation: Levitation practices often involve the manipulation and control of subtle energies within the body. By harnessing and directing these energies, individuals can enhance their overall vitality and well-being. The focused intention and energy manipulation required for levitation can lead to a harmonious flow of energy throughout the body, balancing the various energy centers (chakras), and promoting a state of optimal physical, emotional, and spiritual health.

5.   Self-Discovery and Personal Growth: The pursuit of levitation can be a transformative journey of self-discovery and personal growth. It challenges individuals to push their boundaries, overcome limitations, and explore their potential. The process of developing levitation skills requires dedication, discipline, and perseverance, fostering qualities such as patience, determination, and self-motivation. Through this journey, individuals can develop a greater sense of self-confidence, resilience, and a belief in their own abilities.

6. Emotional Well-being: Levitation practices have the potential to enhance emotional well-being by promoting relaxation, reducing stress, and fostering a sense of inner peace. The focused and meditative nature of levitation exercises allows individuals to enter a state of calm and tranquility, providing a respite from the demands of daily life. Regular practice can help individuals cultivate emotional balance, improve self-awareness, and develop effective stress management techniques.

7. Creative Expression: Levitation can also serve as a form of creative expression. Whether it is performing levitation in a theatrical or artistic context or using levitation as a means to explore new perspectives and ideas, this ability offers a unique medium for self-expression. Levitation can inspire creativity, encourage imaginative thinking, and provide a platform for artistic exploration in various fields, such as dance, photography, and visual arts.

In conclusion, beyond its captivating and awe-inspiring nature, levitation has the potential to offer a range of benefits. From physical fitness and mental well-being to spiritual growth and self-discovery, engaging in levitation practices can be a transformative and enriching experience. Whether pursued for personal development, entertainment, or spiritual purposes, levitation opens up new possibilities and expands our understanding of human potential.

Preparing for Levitation

Levitation, the remarkable ability to defy gravity and float in mid-air, is an extraordinary phenomenon that has captivated the human imagination for centuries. While achieving levitation may seem like a mystical feat, it often requires diligent preparation, both physically and mentally. In this article, we will explore the essential steps to prepare oneself for the practice of levitation.

1.   Physical Conditioning: Physical fitness and conditioning play a crucial role in preparing for levitation. Developing a strong and flexible body is essential for supporting the levitation process. Engaging in regular exercise, particularly exercises that target core strength, balance, and flexibility, can greatly enhance your physical preparedness. Incorporate activities such as yoga, Pilates, strength training, and cardiovascular exercises into your routine to improve your overall fitness level.

2.   Core Strengthening: A strong core is vital for maintaining stability and control during levitation. Focus on exercises that target the abdominal muscles, back muscles, and deep stabilizing muscles of the core. Planks, sit-ups, Russian twists, and yoga poses like Boat Pose and Plank Pose are excellent for strengthening the core. A strong core provides a solid foundation for balance and control during levitation practices.

3.   Flexibility and Stretching: Flexibility is essential for achieving and maintaining different levitation positions. Incorporate regular stretching exercises into your routine to improve your overall flexibility. Pay special attention to stretches that target the muscles in your legs, hips, and lower back, as these areas play a crucial role in levitation. Yoga, stretching routines, and dedicated flexibility exercises can help improve your range of motion and enhance your ability to perform levitation techniques.

4.   Mental Focus and Concentration: Preparing for levitation goes beyond physical conditioning; it also requires mental focus and concentration. Cultivating a calm and focused mind through meditation and mindfulness practices can enhance your ability to achieve levitation. Practice mindfulness exercises to improve your concentration and learn to quiet your mind. Developing mental discipline will help you enter a state of deep focus and concentration during levitation practices.

5.   Visualization Techniques: Visualization is a powerful tool in preparing for levitation. Use your imagination to vividly visualize yourself floating effortlessly in mid-air. Create detailed mental images of the levitation experience, focusing on the sensations, feelings of weightlessness, and control over your body. Regularly practice visualization exercises to strengthen your ability to mentally create the experience of levitation.

6.   Create a Serene Environment: Creating a serene and harmonious environment is essential for successful levitation practices. Find a quiet and peaceful space where you can perform your levitation exercises without distractions. Remove any clutter or objects that may hinder your movements. Consider incorporating soothing music, candles, or incense to create a tranquil atmosphere that promotes relaxation and focus.

7.   Seek Guidance and Training: If you are serious about learning levitation, seeking guidance and training from experienced practitioners can be invaluable. Find reputable teachers or mentors who can provide proper guidance, techniques, and insights into the practice of levitation. They can help you navigate the challenges, offer personalized advice, and ensure that you are practicing safely and effectively.

Remember that levitation is a skill that requires dedication, practice, and patience. It may take time to achieve significant results, but with consistent effort and a positive mindset, progress can be made. Stay committed to your practice, enjoy the journey, and remain open to the possibilities that levitation can bring.

In conclusion, preparing for levitation involves a holistic approach that encompasses physical conditioning, mental focus, and creating an optimal environment. By taking the necessary steps to strengthen your body, cultivate mental discipline, and seek guidance, you can set yourself on a path to explore and experience the extraordinary phenomenon of levitation.

Physical Fitness and Conditioning

Physical fitness and conditioning are essential components of a healthy and active lifestyle. Engaging in regular exercise and maintaining a good level of physical fitness not only benefits your body but also contributes to overall well-being and quality of life. In this article, we will explore the importance of physical fitness and conditioning and provide tips on how to incorporate them into your daily routine.

1. Benefits of Physical Fitness: Physical fitness encompasses various aspects of health, including cardiovascular endurance, muscular strength, flexibility, and body composition. Regular physical activity offers numerous benefits, such as:

• Improved cardiovascular health: Engaging in aerobic exercises, such as jogging, cycling, or swimming, strengthens the heart and improves blood circulation, reducing the risk of heart disease and other cardiovascular conditions.

• Increased muscular strength and endurance: Resistance training, such as weightlifting or bodyweight exercises, helps build muscle strength and endurance, enhancing overall physical performance and reducing the risk of injury.

• Enhanced flexibility and joint mobility: Stretching exercises, yoga, or Pilates improve flexibility, promote better posture, and increase joint range of motion, leading to improved agility and reduced muscle stiffness.

•   Weight management: Regular physical activity helps control body weight by burning calories and increasing metabolic rate, contributing to healthy weight management and reducing the risk of obesity.

•   Improved mental well-being: Physical exercise releases endorphins, also known as "feel-good" hormones, which elevate mood, reduce stress, and promote mental well-being.

•   Enhanced energy levels: Regular physical activity increases energy levels and combats fatigue, improving overall productivity and vitality.

2.   Components of Physical Fitness: Physical fitness comprises several components that work together to support overall health and performance:

•   Cardiovascular endurance: The ability of the heart and lungs to supply oxygen to the muscles during sustained physical activity.

•   Muscular strength: The capacity of muscles to exert force against resistance.

•   Muscular endurance: The ability of muscles to perform repetitive contractions over an extended period.

•   Flexibility: The range of motion of joints and muscles.

•   Body composition: The ratio of fat, muscle, and other tissues in the body.

A well-rounded fitness routine should include exercises that target each of these components to ensure overall physical fitness and conditioning.

3.   Incorporating Physical Fitness into Your Routine: To improve physical fitness and conditioning, consider the following tips:

•   Engage in aerobic activities: Aim for at least 150 minutes of moderate-intensity aerobic exercise or 75 minutes of vigorous-intensity aerobic exercise per week. Choose activities you enjoy, such as brisk walking, running, dancing, or cycling.

•   Include strength training: Incorporate strength training exercises two to three times per week. Use free weights, resistance bands, or bodyweight exercises to target major muscle groups.

•   Prioritize flexibility exercises: Perform stretching exercises or activities like yoga or Pilates to improve flexibility and joint mobility. Stretch major muscle groups before and after exercise sessions.

•   Find enjoyable activities: Engage in physical activities that you find enjoyable and incorporate them into your routine. It could be swimming, hiking, playing a sport, or taking dance classes.

•   Set realistic goals: Set achievable goals that are specific, measurable, attainable, relevant, and time-bound (SMART goals). Monitor your progress and make adjustments as needed.

•   Stay consistent: Consistency is key to reaping the benefits of physical fitness. Aim for regular exercise sessions throughout the week and gradually increase intensity or duration over time.

•   Listen to your body: Pay attention to your body's signals and adjust your exercise routine accordingly. Allow for rest and recovery days to prevent overtraining and reduce the risk of injury.

Remember to consult with a healthcare professional or fitness expert before starting any new exercise program, especially if you have pre-existing health conditions or concerns.

In conclusion, physical fitness and conditioning are vital for overall health and well-being. Engaging in regular physical activity, incorporating aerobic exercises, strength training, and flexibility exercises into your routine, can lead to improved cardiovascular health, increased strength and endurance, enhanced flexibility, and better mental well-being. Prioritize your physical fitness, make it a part of your lifestyle, and enjoy the countless benefits it brings to your life.

Mental Preparation and Focus

Levitation, the ability to defy gravity and float in mid-air, is a captivating phenomenon that has fascinated people for ages. While achieving levitation may seem like a physical feat, mental preparation and focus are equally essential in harnessing this extraordinary ability. In this article, we will explore the significance of mental preparation and focus when it comes to levitation practices.

1.   Cultivating a Calm and Focused Mind: Before attempting levitation, it is crucial to cultivate a calm and focused mind. A cluttered and distracted mind can hinder your ability to concentrate and achieve the necessary mental state for levitation. Engage in practices such as meditation, deep breathing exercises, or mindfulness techniques to quiet your mind, reduce mental chatter, and enter a state of inner calm. These practices enhance your ability to concentrate, maintain focus, and prepare your mind for the intricacies of levitation.

2.   Visualizing the Levitation Experience: Visualization is a powerful tool in mental preparation for levitation. Create a clear mental image of yourself levitating effortlessly in mid-air. Visualize the sensation of weightlessness, the control over your body, and the serenity of the experience. Engage all your senses in the visualization process, immersing yourself in the feeling of levitation. Regularly practice visualization exercises to strengthen your ability to create a detailed and vivid mental image, enhancing your overall readiness for the levitation practice.

3. Developing Concentration Skills: Levitation requires a high level of concentration and focus. Developing concentration skills is paramount to maintaining the necessary mental state throughout the levitation practice. Engage in activities that challenge and improve your concentration, such as solving puzzles, playing memory games, or practicing focused reading. By training your mind to sustain attention on a single task, you enhance your ability to stay focused during the complex and demanding process of levitation.

4. Overcoming Mental Barriers: Mental barriers and self-doubt can impede your progress in levitation. It is essential to address and overcome these barriers through positive affirmations and belief in your own abilities. Cultivate a mindset of self-confidence, resilience, and determination. Remind yourself that levitation is a skill that can be developed with practice and dedication. Surround yourself with supportive individuals who encourage and believe in your journey. By maintaining a positive and empowered mindset, you pave the way for success in your levitation endeavors.

5. Embracing Patience and Persistence: Mental preparation for levitation requires patience and persistence. Levitation is a skill that may take time to master, and progress may come in gradual increments. Embrace the journey with a patient and persevering mindset. Understand that setbacks and challenges are part of the process and view them as opportunities for growth and learning. Stay committed to your practice, remain focused on your goals, and trust in your ability to achieve levitation with dedication and persistence.

6.   Seeking Guidance and Mentorship: Seeking guidance and mentorship from experienced practitioners can greatly support your mental preparation for levitation. Connect with individuals who have expertise in the field of levitation or related practices. Their insights, techniques, and guidance can provide valuable perspectives and help you navigate the mental challenges that arise during the levitation journey. Engage in discussions, workshops, or training programs that offer mentorship and guidance to enhance your mental preparedness for levitation.

In conclusion, mental preparation and focus are vital aspects of levitation practices. By cultivating a calm and focused mind, visualizing the levitation experience, developing concentration skills, overcoming mental barriers, embracing patience and persistence, and seeking guidance, you can enhance your mental readiness for levitation. Remember that levitation is not only a physical feat but also a journey that requires mental discipline, belief in oneself, and a determined mindset. With a prepared and focused mind, you can unlock the remarkable potential of levitation and experience the awe-inspiring sensation of defying gravity.

Creating the Right Environment

Levitation, the ability to float in mid-air, is an extraordinary phenomenon that has captivated the human imagination for centuries. While the practice of levitation involves mastering physical and mental techniques, creating the right environment is equally important for successful levitation experiences. In this article, we will explore the significance of creating the right environment to enhance your levitation practice.

1.   Find a Quiet and Serene Space: To create the right environment for levitation, it is crucial to find a quiet and serene space where you can focus and minimize distractions. Choose a room or an area in your home that is free from external disturbances such as noise or interruptions. This will allow you to concentrate and enter a state of deep relaxation, which is conducive to levitation practices.

2.   Declutter the Space: A cluttered environment can create mental and physical obstacles during levitation practices. Before starting your levitation sessions, declutter the space and remove any unnecessary items or objects that may hinder your movements or create distractions. Clearing the space not only promotes physical freedom but also helps create a sense of clarity and focus.

3.   Lighting and Ambiance: Consider the lighting and ambiance of the space where you practice levitation. Natural lighting can be beneficial, as it creates a soothing and energizing atmosphere. If natural light is limited, use soft and warm artificial lighting to create a calming ambiance. Dim the lights or use candles to create a serene environment that promotes relaxation and concentration.

4.   Aromatherapy and Incense: Engaging your sense of smell can greatly enhance the levitation experience. Consider using aromatherapy or burning incense with calming scents, such as lavender, sandalwood, or frankincense, to create a pleasant and relaxing atmosphere. The aroma can help you enter a state of deep relaxation, enhancing your ability to focus and connect with the levitation practice.

5.   Soundscapes and Music: Sound plays a crucial role in setting the right environment for levitation. Choose soothing soundscapes, gentle instrumental music, or calming meditation tracks to create a peaceful and immersive auditory atmosphere. The right sounds can help you relax, drown out external distractions, and deepen your focus during levitation practices.

6.   Comfortable and Supportive Furniture: Select comfortable and supportive furniture or props that aid your levitation practice. Whether it's a soft mat, a cushion, or a specially designed levitation chair, ensure that your body is adequately supported during the practice. Comfortable furniture allows you to relax and concentrate on the levitation experience without unnecessary physical discomfort or strain.

7.   Personalize the Space: Personalizing the levitation space can create a sense of connection and inspiration. Display objects or images that evoke a sense of wonder and spirituality, such as crystals, artwork, or symbols that hold personal significance. Surrounding yourself with meaningful items can enhance the emotional and spiritual aspects of the levitation practice.

8.   Disconnect from Technology: Levitation is a practice that requires deep focus and inner connection. To create the right environment, disconnect from technology and minimize distractions. Turn off your phone or put it on silent mode, and refrain from checking emails or social media during your levitation sessions. By disconnecting from the digital world, you can fully immerse yourself in the present moment and optimize your levitation practice.

Creating the right environment for levitation is essential to foster a sense of tranquility, focus, and inner connection. By finding a quiet space, decluttering, paying attention to lighting, using aromatherapy, selecting soothing sounds, ensuring comfort, personalizing the space, and disconnecting from technology, you set the stage for a profound levitation experience. Remember, the environment you create can significantly impact your ability to enter a state of deep relaxation and harness the power of levitation.

Developing Levitation Techniques

Levitation, the ability to float or hover above the ground, has long been a subject of fascination and intrigue. While levitation may seem like a supernatural ability, it is a skill that can be developed through consistent practice and the mastery of specific techniques. In this article, we will explore some essential levitation techniques that can help you progress on your levitation journey.

1.   Mind-Body Connection: Developing a strong mind-body connection is fundamental to mastering levitation techniques. Start by practicing mindfulness and body awareness exercises. Pay close attention to the sensations and movements of your body. Cultivate a deep understanding of how your mind and body interact and synchronize. This heightened awareness will lay the foundation for more advanced levitation techniques.

2.   Meditation and Visualization: Meditation is a powerful practice that can aid in the development of levitation skills. Regular meditation helps calm the mind, improve focus, and enhance concentration. During meditation sessions, visualize yourself effortlessly floating above the ground. Engage all your senses, imagining the sensation of weightlessness and freedom. Visualize the levitation process in intricate detail, allowing your mind to create a clear mental image of the experience.

3. Energy Manipulation: Levitation often involves the manipulation of energy within the body. Explore various energy-based practices, such as Qigong or Reiki, to learn how to sense and manipulate subtle energy. These practices help you become more attuned to the energetic aspects of levitation. Focus on circulating and directing energy throughout your body, particularly in areas associated with balance and levitation, such as the core and lower body.

4. Breathing Techniques: Effective control of breath is essential for achieving levitation. Practice deep diaphragmatic breathing to enhance your ability to access a relaxed and focused state. Experiment with breath-holding exercises to develop greater breath control. By becoming proficient in regulating your breath, you can facilitate the flow of energy and increase your stability during levitation attempts.

5. Strengthening Core Muscles: A strong and stable core is crucial for successful levitation. Engage in exercises that target the core muscles, such as planks, yoga poses, and Pilates. Strengthening these muscles improves your ability to maintain balance and control during levitation. Additionally, focus on building leg strength and stability through exercises like squats, lunges, and calf raises, as they provide a solid foundation for levitation practices.

6.   Levitation Techniques: Explore different levitation techniques and find the ones that resonate with you. Some common techniques include:

•   Jump Levitation: Practice jumping from a stationary position and focusing on maintaining your body's height and position in the air for as long as possible. This technique helps develop control and balance.

•   Suspension Levitation: Start by lying on your back and gradually lift one limb at a time, focusing on maintaining it in a suspended position. As you progress, attempt to suspend multiple limbs simultaneously, working towards full-body suspension.

•   Chair Levitation: Sit on a sturdy chair and, using your core and leg muscles, gradually lift your body off the chair while maintaining balance. This technique requires strength, control, and focus.

Remember, levitation techniques require patience and persistence. Progress may come gradually, and it is important to celebrate each milestone along the way.

7.   Seek Guidance and Support: Seeking guidance from experienced practitioners or joining a levitation community can provide valuable insights and support during your levitation journey. Engage in discussions, attend workshops, or seek out mentors who can offer guidance and share their own experiences. Learning from others can help refine your techniques and provide motivation and encouragement.

In conclusion, developing levitation techniques requires a combination of mental focus, physical strength, and energetic awareness. By cultivating a strong mind-body connection, practicing meditation and visualization, manipulating energy, mastering breathing techniques, strengthening core muscles, exploring different levitation techniques, and seeking guidance and support, you can progress on your path to levitation. Remember to approach your practice with patience, consistency, and an open mind, allowing yourself to embrace the extraordinary possibilities of levitation.

Grounding and Centering

Levitation, the ability to float or hover above the ground, is an extraordinary phenomenon that requires a balance between soaring heights and a strong connection to the earth. While the focus of levitation often revolves around defying gravity, grounding and centering practices are equally essential to maintain stability, focus, and overall well-being. In this article, we will explore the importance of grounding and centering in the context of levitation.

1.   What is Grounding? Grounding refers to the process of establishing a solid connection with the earth's energy. It involves anchoring yourself in the present moment and establishing a sense of stability, both physically and energetically. Grounding allows you to remain rooted while exploring the realms of levitation.

2.   The Importance of Grounding in Levitation: Grounding is essential in levitation for several reasons. Firstly, it helps you maintain stability and balance during the practice. Levitation involves defying gravity, which can create a sense of disconnection or imbalance. Grounding acts as a counterbalance, keeping you anchored to the earth and providing a solid foundation for your levitation attempts.

Secondly, grounding helps to channel and balance energy. Levitation involves manipulating energy within your body, and grounding ensures that this energy flows smoothly and is properly integrated. It prevents excessive energy buildup or scattered energy that may hinder your levitation progress.

Finally, grounding promotes mental clarity and focus. Levitation requires a calm and focused mind, and grounding techniques can help quiet the mental chatter and distractions. By grounding yourself, you create a sense of inner stability and presence, enabling you to concentrate on the intricacies of levitation.

3.   Techniques for Grounding and Centering: There are several techniques you can practice to enhance your grounding and centering abilities:

a. Barefoot Connection: Walking barefoot on natural surfaces such as grass, sand, or soil allows direct contact with the earth's energy. This practice helps ground and balance your energy, promoting a sense of stability.

b. Rooting Visualization: Imagine roots extending from the soles of your feet deep into the earth, anchoring you firmly. Visualize these roots absorbing nourishing energy from the earth, providing a solid and stable foundation.

c. Breath Awareness: Focus on your breath, bringing your attention to the sensation of inhaling and exhaling. As you breathe, imagine drawing in grounding energy from the earth with each inhalation and releasing any tension or excess energy with each exhalation.

d. Physical Activity: Engage in activities that promote physical grounding, such as yoga, Tai Chi, or gardening. These practices involve mindful movements, connecting you with your body and the earth's energy.

e. Crystals and Stones: Carry or wear grounding crystals such as hematite, black tourmaline, or smoky quartz. These stones are believed to have grounding properties that help stabilize and balance energy.

f. Nature Connection: Spend time in nature, surrounded by trees, plants, and natural elements. Immerse yourself in the sights, sounds, and scents of the natural world, allowing it to rejuvenate and ground your energy.

4. Finding Your Balance: While grounding is crucial, it is also important to find a balance between grounding and exploring the heights of levitation. Strive to create a harmonious integration of both aspects in your practice. Grounding provides stability and rootedness, while levitation opens up possibilities and expands your awareness. Embrace the dance between the earth and the sky, maintaining a connection to both realms.

In conclusion, grounding and centering are vital elements in the practice of levitation. By grounding yourself, you establish stability, balance your energy, and foster mental clarity. Incorporate grounding techniques into your levitation practice to enhance your overall experience and cultivate a harmonious connection between the earth and the soaring heights of levitation.

Meditation and Visualization

Meditation and visualization are powerful practices that play a significant role in the development and enhancement of levitation abilities. By incorporating these techniques into your levitation practice, you can deepen your focus, strengthen your mind-body connection, and unlock the potential for extraordinary experiences. In this article, we will explore how meditation and visualization can aid you on your levitation journey.

1.   The Power of Meditation: Meditation is a practice that involves training the mind to achieve a state of deep relaxation and heightened awareness. It allows you to quiet the mental chatter, cultivate inner stillness, and tap into your inner resources. Incorporating meditation into your levitation practice is beneficial for various reasons:

a. Enhanced Focus: Levitation requires focused attention and concentration. Regular meditation helps sharpen your focus and enables you to direct your attention to the subtleties of the levitation experience. It improves your ability to remain present in the moment and increases your capacity to sustain concentration for extended periods.

b. Stress Reduction: Levitation practice can be physically and mentally demanding. Meditation acts as a powerful tool for stress reduction, promoting relaxation and calmness. It helps you release tension, alleviate anxiety, and create an inner environment conducive to the exploration of levitation.

c. Mind-Body Connection: Developing a strong mind-body connection is vital for successful levitation. Meditation cultivates awareness of bodily sensations, allowing you to better understand and control subtle movements and energies within your body. This heightened connection enhances your ability to coordinate your movements and align them with the levitation experience.

2. Visualization for Levitation: Visualization is a technique that involves using the power of imagination to create mental images and scenarios. When applied to levitation, visualization can greatly enhance your practice. Here's how visualization can contribute to your levitation journey:

a. Creating Clear Intentions: Visualization enables you to set clear intentions for your levitation practice. Visualize yourself effortlessly floating above the ground, maintaining perfect balance, and experiencing the sensation of weightlessness. This mental imagery establishes a blueprint in your mind, aligning your subconscious with your conscious goals.

b. Building Confidence: Visualization allows you to visualize successful levitation experiences, even before they manifest physically. By repeatedly imagining yourself levitating with ease, grace, and stability, you reinforce positive beliefs and build confidence in your abilities. This confidence can significantly impact your progress and overall experience of levitation.

c. Strengthening Neural Pathways: The brain does not differentiate between vividly imagined experiences and physical experiences. By consistently visualizing levitation, you are strengthening neural pathways associated with the practice. This primes your mind and body for the actual levitation experience, making it easier to translate your visualizations into tangible results.

3.　　How to Incorporate Meditation and Visualization into Levitation Practice: To harness the power of meditation and visualization for levitation, follow these steps:

a. Set Aside Dedicated Time: Allocate specific time for meditation and visualization in your levitation practice. Choose a quiet and comfortable space where you can focus without interruptions.

b. Relaxation Techniques: Begin with relaxation techniques such as deep breathing, progressive muscle relaxation, or guided meditations. These techniques help you relax your body and calm your mind, preparing you for the meditation and visualization practice.

c. Guided Visualization: Use guided visualization recordings or scripts specifically tailored for levitation. These resources can provide step-by-step instructions to help you visualize the levitation process and enhance your experience.

d. Mindfulness Meditation: Practice mindfulness meditation, focusing your attention on the present moment and observing your thoughts and sensations without judgment. This cultivates a state of heightened awareness and increases your ability to sustain focus during levitation practice.

e. Visualization Exercises: Engage in visualization exercises where you vividly imagine yourself levitating. Visualize the sensations, emotions, and physical movements associated with levitation. See yourself floating effortlessly, maintaining balance, and experiencing the joy of weightlessness.

f. Consistency and Patience: Incorporating meditation and visualization into your levitation practice requires consistency and patience. Set aside regular practice sessions and commit to them. Be patient with yourself and trust in the process, allowing the benefits of meditation and visualization to unfold over time.

In conclusion, meditation and visualization are powerful tools that can enhance your levitation practice. Through meditation, you cultivate focus, reduce stress, and develop a strong mind-body connection. Visualization allows you to create clear intentions, build confidence, and strengthen neural pathways associated with levitation. By incorporating these practices into your routine, you can accelerate your progress, deepen your experiences, and unlock the true potential of levitation.

Energy Manipulation

Levitation, the ability to float or hover above the ground, involves more than just defying gravity. It requires a deep understanding and manipulation of subtle energies that flow through our bodies and the universe. Energy manipulation techniques play a crucial role in the practice of levitation, allowing practitioners to harness and direct these energies for the purpose of achieving weightlessness and stability. In this article, we will explore the concept of energy manipulation in relation to levitation and discuss some effective techniques to develop this skill.

1.   Understanding Energy: Energy, in the context of levitation, refers to the subtle life force that flows within and around us. It is often referred to as Qi, Prana, or Vital Energy. According to various energy-based philosophies and practices, this energy can be harnessed, directed, and manipulated to achieve extraordinary feats like levitation.

2.   Sensing Energy: The first step in energy manipulation is developing the ability to sense and perceive subtle energies. Here are some practices to help you become more attuned to energy:

a. Mindful Body Scan: Close your eyes and bring your attention to different parts of your body. Notice any sensations, tingling, warmth, or vibrations. As you become more aware of these subtle sensations, you are tuning in to the energetic aspects of your being.

b. Hand Sensing: Rub your hands together briskly to generate heat. Then, slowly bring your palms closer together without touching. Notice if you can feel a magnetic or tingling sensation between your hands. This exercise helps you become sensitive to the energetic field.

c. Visualization: During meditation or relaxation exercises, visualize energy flowing through your body. Imagine it as a vibrant, glowing light that you can sense and move with your intention. This visualization enhances your connection with the energetic aspect of levitation.

3.   Circulating and Directing Energy: Once you can sense energy, the next step is to learn how to circulate and direct it. Here are some techniques to help you accomplish this:

a. Qigong: Qigong is a Chinese energy practice that involves specific movements, postures, and breath control to cultivate and circulate energy. Explore Qigong exercises that focus on cultivating and moving energy through your body, particularly in the lower abdomen and the areas associated with balance and stability.

b. Reiki and Healing Modalities: Reiki and other energy healing modalities can provide valuable insights and techniques for manipulating energy. Seek out experienced practitioners or explore books and resources on energy healing to learn techniques that can be adapted for levitation.

c. Breathwork: Breath is a powerful tool for energy manipulation. Practice specific breathing techniques such as deep diaphragmatic breathing, alternate nostril breathing, or breath retention exercises. These techniques can help you control and direct the flow of energy within your body.

d. Visualization and Intention: Combine visualization with intention to guide the movement of energy. During meditation or levitation practice, visualize energy flowing through specific pathways in your body, following your intended direction. Use your mind's focus and intention to guide the energy to the areas needed for levitation.

4.   Balancing and Harmonizing Energy: Balance and harmony are essential in energy manipulation for levitation. Here are some practices to help you achieve balance:

a. Chakra Balancing: Explore practices that focus on balancing and aligning the chakras, which are energy centers in the body. When the chakras are in balance, the energy flows smoothly, promoting stability and equilibrium.

b. Grounding and Centering: As discussed in previous articles, grounding and centering practices help establish a strong connection with the earth's energy. Grounding techniques, such as walking barefoot on the earth or visualizing roots connecting you to the ground, can assist in stabilizing your energy for levitation.

c. Meditation and Mindfulness: Regular meditation and mindfulness practices help maintain overall balance and equanimity. These practices cultivate a calm and focused state of mind, enabling you to channel and manipulate energy with greater clarity and effectiveness.

Remember that energy manipulation for levitation requires dedication, practice, and a deep understanding of your own energetic system. As you develop your skills in energy manipulation, continue to refine and expand your knowledge through exploration, study, and guidance from experienced practitioners. With patience and consistent practice, you can unlock the power of energy manipulation to enhance your levitation abilities.

Exercises for Beginners

Levitation, the ability to float or hover above the ground, is an awe-inspiring phenomenon that has captivated the imagination of many. While levitation may seem like an extraordinary feat reserved for a select few, there are exercises and techniques that can help beginners develop and enhance their levitation abilities. In this article, we will explore some introductory levitation exercises that can lay the foundation for your levitation practice.

1.   Mindful Body Awareness: Before delving into specific levitation techniques, it's important to cultivate a strong mind-body connection and develop heightened awareness of your physical sensations. The following exercise will help you establish a solid foundation for your levitation practice:

•   Find a quiet and comfortable space to sit or lie down. Close your eyes and bring your attention to your body.

•   Scan your body from head to toe, noticing any areas of tension, sensations, or points of contact with the surface beneath you.

•   Take slow, deep breaths, allowing your breath to flow naturally and effortlessly.

•   With each breath, imagine releasing any tension or heaviness in your body, allowing yourself to feel light and relaxed.

•   As you continue breathing and scanning your body, observe any subtle shifts or sensations that arise. Cultivate a sense of curiosity and openness to the energy within your body.

2.   Visualization Exercises: Visualization is a powerful tool in the practice of levitation. It allows you to create clear mental images and intentions that can influence your energy and movements. The following visualization exercises will help you develop your visualization skills:

•   Sit or lie down in a comfortable position and close your eyes.

•   Visualize yourself floating above the ground, effortlessly and with grace. Imagine the sensation of weightlessness and freedom.

•   Envision a beam of light extending from the core of the earth, passing through your body, and extending upwards towards the sky. Feel the support and stability provided by this connection to the earth's energy.

•   As you visualize yourself levitating, engage all your senses. Imagine the feeling of air beneath you, the gentle breeze on your skin, and the sounds that accompany your floating experience.

•   Repeat this visualization exercise regularly, allowing your mind to create vivid and detailed images of levitation.

3.   Core Strengthening Exercises: Building strength in your core muscles is crucial for maintaining stability and control during levitation. The following exercises will help strengthen your core:

•   Plank: Begin in a push-up position, with your palms on the floor directly beneath your shoulders. Keep your body in a straight line from head to toe, engaging your core muscles. Hold this position for as long as you can, aiming to increase your endurance over time.

•   Leg Raises: Lie on your back with your arms by your sides. Lift your legs off the ground, keeping them straight. Slowly lower your legs back down without touching the ground. Repeat this exercise for several repetitions.

•   Bicycle Crunches: Lie on your back with your hands behind your head. Lift your legs off the ground and bring your left elbow to your right knee while extending your left leg. Alternate sides in a cycling motion while engaging your core muscles.

4.   Breath Control: Mastering breath control is essential for maintaining focus and stability during levitation. Practice the following breathing exercise to enhance your breath control:

•   Sit in a comfortable position with your spine straight. Take a slow, deep breath in through your nose, allowing your abdomen to expand. Exhale slowly through your mouth, drawing your navel towards your spine to engage your core. Repeat this deep breathing exercise for several minutes, focusing on smooth and controlled inhalations and exhalations.

5.   Seek Guidance and Support: As a beginner in levitation, it can be beneficial to seek guidance and support from experienced practitioners or teachers who can provide personalized instruction and feedback. They can offer insights, answer your questions, and help you refine your levitation practice.

Remember, progress in levitation takes time and patience. Be consistent with your practice, and celebrate even the smallest achievements along the way. With dedication and perseverance, you can gradually develop your levitation abilities and embark on an extraordinary journey of self-discovery and exploration.

Floating Feather Technique

The Floating Feather Technique is a well-known exercise in the realm of levitation. It focuses on developing the ability to manipulate subtle energies and create a sense of weightlessness. This technique serves as an excellent starting point for beginners who are eager to explore the fascinating world of levitation. In this article, we will delve into the Floating Feather Technique and guide you through the steps to incorporate it into your levitation practice.

1.   Understanding the Floating Feather Technique: The Floating Feather Technique involves using your focused intention and energy manipulation to lift and control a feather or a lightweight object. By directing your energy towards the feather, you create an energetic connection that allows you to influence its movement and defy gravity. This exercise enhances your ability to perceive and manipulate subtle energies, paving the way for further exploration in levitation.

2.   Preparation: To begin the Floating Feather Technique, you will need a feather or a lightweight object that can be easily affected by subtle movements. Find a quiet and calm space where you can focus without distractions. Take a few moments to ground yourself by connecting with the earth's energy and centering your awareness.

3.   Calming the Mind: Before attempting to manipulate the feather, it is crucial to calm the mind and cultivate a state of focused awareness. You can achieve this through various techniques such as deep breathing, meditation, or visualization. Clear your mind of any distractions and bring your attention to the present moment.

4.   Creating an Energetic Connection: Hold the feather in your hand, gently feeling its texture and weight. Close your eyes and visualize a subtle energy flowing from your core into your hand and extending to the feather. Imagine a connection between your energy and the energy of the feather, as if they are intertwined.

5.   Focused Intention: With your eyes still closed, set a clear intention to lift the feather with your energy. Visualize the feather becoming weightless, defying gravity, and floating in the air. Focus your attention on this intention, allowing it to permeate your being.

6.   Directing the Energy: Using your focused intention, gradually increase the flow of energy from your hand to the feather. Imagine your energy gently lifting the feather, allowing it to rise slowly and gracefully. Be patient and maintain your focus as you guide the feather with your energy.

7.   Observing the Feather: As the feather responds to your energy, observe its movements closely. Notice any shifts in direction, height, or speed. Use your intention to direct its movement, experimenting with different ways of manipulating the feather through your energy.

8. Practice and Progression: Consistency is key when practicing the Floating Feather Technique. Set aside dedicated time each day to engage in this exercise. With regular practice, you will gradually develop greater control and finesse in manipulating the feather. As you become more proficient, you can explore variations such as levitating multiple feathers or larger lightweight objects.

9. Patience and Persistence: Levitation is a skill that requires patience and persistence. Be gentle with yourself as you navigate the learning process. It may take time to achieve significant results, but every small accomplishment is a step forward in your levitation journey.

10. Seek Guidance and Support: If you encounter challenges or have questions along the way, consider seeking guidance and support from experienced practitioners or teachers. They can provide valuable insights, feedback, and encouragement to help you refine your technique.

The Floating Feather Technique is a wonderful exercise to develop your energy manipulation skills and lay the foundation for further exploration in levitation. Embrace the joy of the process, stay open to new experiences, and enjoy the remarkable journey of levitation that lies ahead.

Balloon Lift Exercise

The Balloon Lift Exercise is a playful and effective technique that can help beginners develop their levitation abilities. It involves using focused intention and energy manipulation to lift and control a balloon. This exercise allows you to experience the sensation of defying gravity and serves as an excellent introduction to the world of levitation. In this article, we will explore the Balloon Lift Exercise and guide you through the steps to incorporate it into your levitation practice.

1.   Preparation: To begin the Balloon Lift Exercise, you will need a helium-filled balloon and a quiet space where you can focus without distractions. Choose a balloon that is large enough to be easily affected by your movements but not too heavy to handle.

2.   Centering and Grounding: Before attempting the Balloon Lift Exercise, take a few moments to center yourself and connect with the earth's energy. Stand or sit comfortably, close your eyes, and take deep breaths. Visualize roots extending from the soles of your feet, grounding you to the earth. Feel the stability and support of the earth beneath you.

3.   Holding the Balloon: Hold the string of the helium balloon in your hand. Gently feel the lightness of the balloon and establish a connection with it. Close your eyes and focus your attention on the balloon, allowing yourself to become fully present in the moment.

4.   Focused Intention: Set a clear intention to lift and control the balloon using your energy. Visualize the balloon becoming weightless and imagine yourself having the ability to guide its movements with your focused intention.

5.   Directing Your Energy: With your eyes closed, visualize a flow of energy from your core extending through your arm and into the balloon. Imagine your energy connecting with the balloon, creating an energetic link between you and the object. Feel the energetic connection between your energy and the balloon's energy.

6.   Lifting the Balloon: Using your focused intention and energy, gradually increase the flow of energy into the balloon. Visualize the balloon gently rising and defying gravity as you guide its movement. Be patient and allow the balloon to respond to your energy.

7.   Controlling the Balloon: Once the balloon is airborne, practice controlling its movement. Experiment with different energy manipulations to guide the balloon in various directions. Use your intention and energy to maneuver the balloon up, down, and sideways.

8.   Sensing the Energy: As you lift and control the balloon, pay attention to the subtle sensations and feedback you receive. Notice the energetic connection between you and the balloon. Tune in to any shifts in energy, vibrations, or changes in your own body sensations as you interact with the balloon.

9. Play and Explore: Enjoy the playful nature of the Balloon Lift Exercise. Allow yourself to experiment and explore different ways of manipulating the balloon with your energy. Embrace the joy and wonder of levitating the balloon, and let your imagination soar as you discover new possibilities.

10. Reflect and Learn: After each session of the Balloon Lift Exercise, take a few moments to reflect on your experience. Consider the sensations, emotions, and insights you gained during the exercise. Note any progress you have made and areas where you would like to further develop your skills.

Remember, levitation is a skill that requires practice, patience, and a deep connection with your energy. Embrace the Balloon Lift Exercise as a playful opportunity to explore and enhance your levitation abilities. With dedication and consistent practice, you can expand your understanding of levitation and unlock the extraordinary potential within you.

Levitation Cushion Practice: Enhancing Your Levitation Abilities

The Levitation Cushion Practice is a unique and effective technique that can assist in the development of levitation abilities. It involves using a specially designed cushion to create a supportive and conducive environment for your levitation practice. This article will explore the concept of Levitation Cushion Practice and guide you through the steps to incorporate it into your levitation training.

1.   Understanding the Levitation Cushion: A Levitation Cushion is a cushion specifically designed to provide comfort and support during levitation practice. It is usually made of soft yet durable materials and is shaped to accommodate different sitting or lying positions. The cushion serves as a tool to enhance your focus, stability, and energetic connection while practicing levitation.

2.   Selecting the Right Cushion: Choose a Levitation Cushion that suits your comfort and preferences. Consider the size, shape, and material of the cushion. Some cushions are designed specifically for meditation, while others have additional features such as adjustable height or ergonomic support. Find a cushion that allows you to maintain a relaxed and upright posture during your levitation practice.

3.   Creating a Sacred Space: Before starting your Levitation Cushion Practice, create a sacred space that is free from distractions. Choose a quiet area where you can focus without interruptions. Clear the space of clutter and create an ambiance that promotes relaxation and tranquility. You may incorporate elements such as candles, incense, or soothing music to enhance the atmosphere.

4.   Setting Intentions: Before sitting on the Levitation Cushion, take a moment to set clear intentions for your practice. Reflect on your goals and aspirations in developing your levitation abilities. Visualize yourself effortlessly floating or levitating, and hold that image in your mind as you prepare for the practice.

5.   Proper Alignment and Posture: Sit or lie down on the Levitation Cushion with proper alignment and posture. Keep your spine straight, allowing the energy to flow freely through your body. Relax your shoulders, soften your facial muscles, and find a comfortable position that enables you to maintain balance and stability.

6.   Breath Awareness: Focus your attention on your breath. Take slow, deep breaths, allowing your breath to flow naturally and effortlessly. Notice the sensation of each inhalation and exhalation, bringing your awareness to the present moment. Use your breath as an anchor to cultivate mindfulness and concentration.

7.   Energetic Connection: With your body relaxed and your mind centered, direct your attention to establishing an energetic connection with the Levitation Cushion. Visualize an energetic link between you and the cushion, as if it is an extension of your own energy field. Feel the support and grounding energy of the cushion beneath you.

8.   Energy Manipulation: Engage in energy manipulation techniques to enhance your levitation practice. Visualize the flow of energy within your body and direct it towards the desired outcome of levitation. Use your intention and focused energy to create a sense of lightness and lift.

9.   Exploration and Practice: Once you have established a connection with the Levitation Cushion and directed your energy, begin to explore subtle movements and shifts. Experiment with different energetic techniques to gradually lift your body or create a sense of weightlessness. Be patient and allow yourself to become familiar with the sensations and possibilities that arise.

10.   Reflect and Integrate: After each session of Levitation Cushion Practice, take a few moments to reflect on your experience. Notice any shifts in your energy, physical sensations, or insights gained during the practice. Integrate these experiences into your understanding of levitation and use them as stepping stones for further exploration and growth.

The Levitation Cushion Practice offers a supportive and focused environment for enhancing your levitation abilities. By incorporating this practice into your training routine, you can cultivate stability, concentration, and a deeper connection with your energetic self. Embrace the possibilities that the Levitation Cushion Practice brings, and enjoy the journey of unlocking your extraordinary potential for levitation.

Advanced Levitation Techniques: Unlocking the Extraordinary

Levitation has long captivated the human imagination, representing a state of heightened consciousness and an extraordinary connection with energy. As you progress in your levitation practice, you may find yourself seeking new challenges and wanting to explore advanced techniques. In this article, we will delve into some advanced levitation techniques that can take your practice to new heights.

1.   Levitation through Mind Power: One of the most advanced techniques in levitation involves harnessing the power of the mind to lift and suspend the body. This technique requires deep concentration, focused intention, and a strong connection with your energy field. Through mental discipline and visualization, you can project your energy downward to create an upward force, defying gravity and levitating your body.

2.   Levitation with Sound and Vibrations: Sound and vibrations have a profound impact on energy and consciousness. Advanced levitation techniques incorporate the use of sound frequencies, chanting, or specific tones to induce altered states of consciousness and facilitate levitation. By resonating with the vibrations around you, you can harmonize your energy and create a conducive environment for levitation.

3.   Levitation through Astral Projection: Astral projection is the phenomenon of consciously separating your consciousness from the physical body and traveling in a non-physical realm. Advanced practitioners of levitation can combine the practice of astral projection with levitation, allowing their astral body to levitate and explore higher dimensions. This technique requires mastery of astral projection and a deep understanding of energy manipulation.

4.   Levitation with Sacred Geometry: Sacred geometry, the study of geometric patterns found in nature and spirituality, holds immense power in energy manipulation. Advanced levitation techniques utilize sacred geometry symbols, such as the Merkaba or Sri Yantra, to amplify and direct energy for levitation. By meditating on these symbols and aligning with their energetic frequencies, practitioners can access heightened states of consciousness and levitate.

5.   Levitation through Quantum Field Interaction: The principles of quantum physics suggest that reality is influenced by the observer's consciousness and intention. Advanced levitation techniques incorporate an understanding of quantum field interaction to manifest levitation. By aligning your consciousness with the quantum field and directing your intention, you can manipulate the fabric of reality and levitate your physical body.

6. Levitation with Group Energy: Group energy amplifies individual intentions and capabilities. Advanced levitation techniques involve practicing levitation in a group setting, where collective energy is harnessed and focused towards the common goal of levitation. This technique taps into the power of collective consciousness, enhancing the potential for achieving profound levitation experiences.

7. Levitation through Energetic Mastery: Advanced levitation requires mastery over the subtle energy systems within the body. Techniques such as chakra activation, energy channeling, and energy manipulation play a crucial role in achieving levitation. By refining your energetic awareness and cultivating a harmonious flow of energy, you can elevate your levitation abilities to new levels.

8. Levitation with Higher Consciousness States: States of expanded consciousness, such as deep meditation or transcendental states, can act as gateways to advanced levitation experiences. By accessing higher realms of consciousness, practitioners can tap into the unlimited potential of their being and effortlessly levitate. These states require dedication, practice, and a willingness to surrender to the flow of universal energy.

Remember, advanced levitation techniques should be approached with respect, humility, and a solid foundation in the fundamental principles of levitation. It is essential to continue honing your skills, maintaining a regular practice, and seeking guidance from experienced teachers or practitioners who can provide guidance and support on your journey.

As you delve into advanced levitation techniques, remain open to the transformative power of these practices. Embrace the unknown, trust in your innate abilities, and allow yourself to explore the extraordinary realm of levitation.

Levitating Objects: Unveiling the Power of Telekinesis

The ability to levitate objects has fascinated and intrigued humanity for centuries. The notion of moving things with the power of the mind alone, known as telekinesis, has captivated our imaginations. While telekinesis may seem like a concept from the realm of science fiction, many believe that it is a real and attainable ability. In this article, we will explore the idea of levitating objects and delve into the techniques and practices associated with telekinesis.

1. Understanding Telekinesis: Telekinesis is the psychic ability to manipulate objects through the power of the mind. It involves harnessing and directing energy to influence the physical world. While the scientific community may still be skeptical about its existence, numerous anecdotal accounts and experiences suggest that telekinesis is indeed possible.

2. Developing Mental Focus and Concentration: To begin your journey into levitating objects, it is crucial to cultivate mental focus and concentration. Practice meditation and visualization exercises to enhance your ability to quiet the mind and direct your attention with precision. Strengthening your mental focus will serve as a foundation for your telekinetic endeavors.

3.   Energy Manipulation: Telekinesis is closely tied to energy manipulation. By developing an awareness of your own energy and learning to control and direct it, you can increase your telekinetic potential. Engage in energy exercises, such as working with chi or prana, to deepen your understanding and mastery of energy manipulation.

4.   Establishing a Connection with Objects: To levitate an object, it is essential to establish a strong energetic connection with it. Spend time with the object, hold it in your hands, and visualize an energetic link between yourself and the item. Develop a sense of familiarity and intimacy with the object, as if it is an extension of your own energy field.

5.   Start with Small and Lightweight Objects: When beginning your practice of telekinesis, it is advisable to start with small and lightweight objects. Feather, a small piece of paper, or a lightweight ball can serve as suitable objects to practice with. These objects are more responsive to subtle energetic influences, making them ideal for beginners.

6.   Visualization and Intention: Visualization plays a crucial role in telekinesis. Imagine the object being lifted by an invisible force, defying gravity. Visualize the object floating in the air, suspended by your focused intention. Allow yourself to deeply believe in your ability to move the object through the power of your mind.

7. Patience and Persistence: Developing telekinetic abilities takes time, patience, and persistence. It is important to approach the practice with a sense of curiosity and playfulness, rather than placing undue pressure on yourself. Celebrate even the smallest progress and maintain a consistent practice routine to strengthen your telekinetic skills.

8. Ethical Considerations: As you explore telekinesis and levitating objects, it is crucial to emphasize ethical considerations. Respect the autonomy and free will of objects and individuals. Use your telekinetic abilities responsibly, with the intention of growth, understanding, and harmony.

9. Seek Guidance and Support: Embarking on the journey of telekinesis can be both exciting and challenging. It is beneficial to seek guidance and support from experienced practitioners or mentors who can provide insights, techniques, and guidance along the way. Joining communities or groups interested in psychic phenomena can also be helpful for sharing experiences and learning from others.

10. Embrace the Journey: Remember that telekinesis is a personal journey of self-discovery and exploration. Embrace the process, and let go of expectations and limitations. Each step you take in developing your telekinetic abilities brings you closer to unlocking the extraordinary power within yourself.

Levitating objects through telekinesis is a fascinating and awe-inspiring ability that demonstrates the incredible potential of the human mind. With dedication, practice, and an open mind, you can embark on a remarkable journey of self-discovery and unfold the mysteries of telekinesis.

Levitating Self: Exploring the Heights of Personal Transcendence

Levitating oneself, the pinnacle of levitation mastery, represents the ultimate manifestation of mind over matter. The ability to defy gravity and elevate one's physical body through focused intention and energy manipulation is a profound and transformative experience. In this article, we will delve into the concept of levitating oneself and explore the practices and techniques that can assist in achieving this remarkable feat.

1.   The Power of Mind and Belief: Levitating oneself begins with the power of the mind and unwavering belief in the extraordinary capabilities of the human spirit. Recognize that the mind plays a significant role in shaping our reality, and by cultivating a strong belief in your ability to levitate, you open the door to its manifestation.

2.   Mastery of Energy Manipulation: Energy manipulation is at the core of levitating oneself. Develop a deep understanding of your own energy field and learn to channel and direct it with precision. Practice energy exercises, such as Qi Gong or Reiki, to enhance your energetic awareness and control. By harnessing and directing this energy, you can create the necessary conditions for self-levitation.

3. Cultivating Physical and Mental Alignment: Achieving levitation requires a state of profound physical and mental alignment. Engage in practices that promote physical fitness, flexibility, and strength. Incorporate activities such as yoga, tai chi, or dance to enhance body awareness and control. Additionally, cultivate mental focus and clarity through meditation, visualization, and mindfulness techniques.

4. Connecting with Higher Consciousness: Levitating oneself often involves tapping into higher states of consciousness. Explore transcendental meditation, deep contemplation, or other spiritual practices to access heightened levels of awareness. By expanding your consciousness and connecting with the universal energy, you can elevate your ability to levitate.

5. Letting Go of Limiting Beliefs and Fear: Limiting beliefs and fear can hinder your progress in levitating oneself. Release any doubts, skepticism, or fear of failure that may arise. Embrace a mindset of unlimited possibilities and trust in your inherent ability to transcend the limitations of the physical realm.

6. Progressive Training and Incremental Progress: Levitating oneself is a skill that requires progressive training and patience. Start by practicing small increments of lift-off, gradually increasing the duration and height of levitation over time. Celebrate each milestone along the way, and maintain a consistent practice routine to further develop your levitation abilities.

7.   Seek Guidance and Mentorship: Embarking on the journey of levitating oneself can be both exhilarating and challenging. Seek guidance and mentorship from experienced practitioners or teachers who can provide valuable insights, techniques, and support. Their expertise and guidance can assist you in overcoming obstacles and refining your levitation skills.

8.   Integrate Mind, Body, and Spirit: Levitating oneself is not solely a physical feat but a harmonious integration of mind, body, and spirit. Cultivate a holistic approach to your practice by nurturing your physical health, sharpening your mental focus, and nourishing your spiritual well-being. Embrace practices that promote balance, such as meditation, breathwork, and self-reflection.

9.   Surrender to the Flow: While focused intention and energy manipulation are essential, it is equally crucial to surrender to the flow of the levitation experience. Trust in the wisdom of your higher self and allow the levitation to unfold naturally. Embrace the sensation of weightlessness, and let go of the need to control or force the process.

10.   Embrace the Transcendence: Levitating oneself is not only a physical accomplishment but also a journey of personal transcendence. Embrace the profound transformation that comes with this practice. Allow it to expand your consciousness, deepen your spiritual connection, and open new dimensions of self-discovery and empowerment.

Levitating oneself represents the ultimate expression of human potential, pushing the boundaries of what we perceive as possible. Approach this practice with reverence, dedication, and an open heart, as it has the power to unlock profound states of transcendence and liberation.

Levitation in Motion: Transcending the Boundaries of Movement

Levitation, the ability to defy gravity and float above the ground, is often associated with stillness and serenity. However, there is another dimension to levitation that explores the possibilities of movement while suspended in mid-air. Levitation in motion is a captivating and dynamic practice that pushes the boundaries of what is considered possible. In this article, we will explore the concept of levitation in motion and the incredible potential it holds.

1.   The Fusion of Levitation and Movement: Levitation in motion combines the freedom of movement with the weightlessness of levitation. It allows practitioners to transcend the limitations of gravity and explore new dimensions of expression. By integrating levitation into movement practices such as dance, martial arts, or acrobatics, a new realm of artistic and physical possibilities unfolds.

2.   Harnessing Energy and Momentum: To achieve levitation in motion, it is crucial to harness energy and momentum. Practitioners utilize their understanding of energy manipulation to direct and channel forces in their movements. By aligning the body, breath, and intention, they create a harmonious flow of energy that propels them into levitated motion.

3.   Engaging the Core and Balance: Maintaining a strong core and balance is essential in levitation in motion. The core serves as the energetic center, providing stability and control. Developing core strength through targeted exercises such as yoga or Pilates enables practitioners to maintain balance while in levitated motion.

4.   Expanding Spatial Awareness: Levitating in motion requires a heightened spatial awareness. Practitioners must attune themselves to their surroundings, anticipate movements, and adjust their energy accordingly. This expansion of awareness enhances the overall fluidity, precision, and grace of their levitated movements.

5.   Embracing Creativity and Expression: Levitation in motion is a platform for creative expression. It offers an opportunity to explore unique and unconventional movement patterns. Practitioners can experiment with different styles, rhythms, and dynamics, weaving their personal artistry into the levitated motion. It becomes a form of self-expression and a means to communicate emotions and stories.

6.   Integrating Flow and Transitions: Flow and seamless transitions are key elements in levitation in motion. Smoothly transitioning from one levitated movement to another requires mastery of control, energy modulation, and synchronization. Practitioners aim to create a continuous and fluid sequence of movements that appear effortless and connected.

7.   Embodying Grace and Lightness: Levitation in motion exudes an aura of grace and lightness. As practitioners float through the air, their movements become imbued with a sense of weightlessness and ethereality. Cultivating a sense of inner lightness and embodying grace allows for a more authentic and captivating levitation experience.

8.   Mind-Body Connection and Presence: The practice of levitation in motion deepens the mind-body connection. Practitioners must be fully present in the moment, attuning their minds to the subtleties of their bodies and the surrounding energy. By cultivating mindfulness and presence, they can fully embrace the unique experience of levitation in motion.

9.   Collaborative Possibilities: Levitation in motion also opens up collaborative possibilities. Practitioners can engage in partner or group levitation experiences, synchronizing their movements and energies to create breathtaking displays of levitated motion. Collaborative exploration enhances creativity, trust, and connection with fellow practitioners.

10.   Embracing the Joy of Levitated Motion: Ultimately, levitation in motion is a joyful and liberating practice. It allows practitioners to experience a sense of freedom and transcendence as they move through the air. Embrace the joy and childlike wonder that comes with levitated motion, and let it fuel your passion for exploration and self-expression.

Levitation in motion offers a thrilling and boundary-pushing journey for those willing to explore the marriage of movement and levitation. It invites practitioners to go beyond the limitations of gravity, celebrate their unique artistry, and unlock new realms of physical and creative expression.

Overcoming Challenges on the Path to Mastery

The path to mastering levitation is a journey filled with wonder, excitement, and growth. However, like any skill or practice, it comes with its fair share of challenges. In this article, we will explore some of the common obstacles faced by individuals on the quest for levitation and provide guidance on how to overcome them.

1.   Patience and Persistence: One of the primary challenges in levitation is cultivating patience and persistence. Levitation is not an overnight accomplishment but a skill that requires consistent practice and dedication. Embrace the process and understand that progress may be gradual. Celebrate even the smallest achievements along the way to stay motivated and persistent.

2.   Mind Over Matter: Levitation is as much a mental feat as it is a physical one. It requires mastery over the mind and the ability to transcend limiting beliefs and self-doubt. Overcoming the notion that levitation is impossible or reserved for a select few is crucial. Cultivate a mindset of possibility and believe in your own potential to achieve levitation.

3.   Developing Focus and Concentration: Maintaining focus and concentration is vital in levitation. Distractions and wandering thoughts can disrupt the energetic flow necessary for successful levitation. Practice meditation, mindfulness, and visualization techniques to sharpen your focus and strengthen your ability to hold your attention on the task at hand.

4.   Managing Energy: Energy management is key in levitation. Learning to harness and direct energy effectively can greatly impact your levitation practice. However, balancing and harmonizing your energy can be challenging. Explore different energy practices such as qigong or yoga to develop a greater awareness of your energy flow and to cultivate the ability to modulate it as needed.

5.   Overcoming Physical Limitations: Physical limitations can pose challenges on the path to levitation. Flexibility, strength, and overall fitness play a role in achieving and maintaining levitated states. Incorporate exercises that enhance your physical abilities, such as yoga, Pilates, or strength training, to overcome these limitations. Gradually push your boundaries and work towards improving your physical capacity.

6.   Seeking Guidance and Support: Embarking on the journey of levitation can feel daunting at times. Seeking guidance and support from experienced practitioners or mentors can be invaluable. They can provide insights, techniques, and encouragement to help you navigate challenges and accelerate your progress. Joining a community or attending workshops and retreats dedicated to levitation can also offer a supportive network.

7.   Embracing Failures as Learning Opportunities: Failure is an inherent part of any learning process, including levitation. Instead of becoming discouraged by setbacks or unsuccessful attempts, embrace them as valuable learning opportunities. Analyze what went wrong, adjust your approach, and keep pushing forward. Each failure brings you closer to understanding the nuances of levitation and improving your technique.

8. Trusting the Process: Trust in the process of levitation and surrender to its mysteries. Understand that progress may unfold in unexpected ways and at different rates for each individual. Avoid comparing your journey to others and trust in your own unique path.

Embracing Doubt and Skepticism on the Path to Discovery

The practice of levitation often elicits doubt and skepticism from both ourselves and others. The concept of defying gravity challenges our understanding of the physical world. In this article, we will explore how to deal with doubt and skepticism when pursuing levitation, and how to stay committed to our journey of discovery.

1.   Validate Your Experience: When faced with doubt and skepticism, it is essential to validate your own experiences. Trust in the moments of levitation you have witnessed or felt, no matter how brief or subtle they may be. Recognize that your personal experiences hold value and should not be dismissed simply because they defy conventional belief systems.

2.   Educate Yourself: Knowledge is a powerful tool in overcoming doubt and skepticism. Dive deep into the study of levitation, exploring the historical accounts, scientific research, and spiritual teachings surrounding this phenomenon. By expanding your understanding and knowledge base, you can strengthen your own convictions and confidently navigate discussions with skeptics.

3.   Surround Yourself with Supportive Communities: Connecting with like-minded individuals who share your interest in levitation can provide valuable support. Seek out communities, forums, or groups where you can share your experiences, exchange knowledge, and find encouragement. Being part of a supportive community can help you maintain confidence and reinforce your belief in the possibilities of levitation.

4.   Engage in Objective Self-Reflection: Self-reflection is a powerful practice in overcoming doubt and skepticism. Take the time to reflect on your motivations, intentions, and experiences. Assess whether doubt arises from external influences or internal insecurities. By examining your own beliefs and understanding, you can develop a stronger foundation and address any lingering doubts or skepticism within yourself.

5.   Embrace Skepticism as a Catalyst for Growth: Skepticism, when approached with an open mind, can serve as a catalyst for growth. Engage in constructive conversations with skeptics, listen to their perspectives, and understand their skepticism. Use these interactions as an opportunity to refine your own understanding, challenge your beliefs, and strengthen your arguments based on evidence and personal experiences.

6.   Seek Scientific Exploration: Scientific exploration and research can bridge the gap between skepticism and belief. Look for scientific studies or institutions that have delved into the investigation of levitation. Engaging with scientific explanations and evidence can provide a balanced perspective and help you articulate the potential mechanisms behind levitation.

7.   Cultivate Inner Confidence: Developing inner confidence is key in dealing with doubt and skepticism. Believe in your own experiences and trust in your abilities. Engage in daily affirmations, visualization techniques, or mindfulness practices that reinforce your self-belief. By cultivating inner confidence, you can navigate external skepticism with resilience and clarity.

8.   Set Boundaries and Protect Your Energy: Not all doubt and skepticism come from a place of healthy skepticism. Some individuals may project their own limiting beliefs or fears onto you. Set clear boundaries and protect your energy by limiting interactions with those who consistently undermine your experiences or discourage your journey. Surround yourself with supportive individuals who uplift and inspire you.

9.   Focus on Personal Growth: Ultimately, the journey of levitation is about personal growth and self-discovery. Instead of seeking validation or approval from others, focus on your own progress and development. Embrace the transformative power of your experiences and allow them to shape your understanding of yourself and the world around you.

10.   Stay Open to Possibilities: Remaining open-minded is crucial when dealing with doubt and skepticism. While skepticism challenges us, staying open to possibilities allows us to explore new realms of understanding. Embrace curiosity and continue to explore the practice of levitation with an open heart and mind. This attitude of openness can lead to profound discoveries and a deeper appreciation for the mysteries of the universe.

In conclusion, dealing with doubt and skepticism is an integral part of the levitation journey. By validating your experiences, educating yourself, seeking support, engaging in self-reflection, and staying open-minded, you can navigate the challenges and maintain your commitment to the exploration of levitation. Embrace the transformative power of doubt and skepticism, and let them fuel your growth and determination on this extraordinary path of discovery.

Managing Energy Drain for Sustainable Practice

The practice of levitation can be a transformative and exhilarating experience. However, it also demands a significant amount of energy, both physically and energetically. Managing energy drain is crucial for sustaining a consistent and balanced levitation practice. In this article, we will explore effective strategies for managing energy drain and optimizing your levitation journey.

1.   Prioritize Self-Care: Self-care plays a fundamental role in managing energy drain. Prioritize rest, proper nutrition, and adequate hydration to support your physical and energetic well-being. Give yourself permission to rest when needed, as fatigue can hinder your ability to access and maintain levitated states. Engage in activities that rejuvenate and nourish your body, mind, and spirit.

2.   Develop a Grounding Practice: Grounding is essential for maintaining a balanced energy flow. When engaging in levitation, it is easy to become ungrounded and lose connection with the Earth's energy. Incorporate grounding practices such as walking barefoot on the earth, practicing yoga, or meditation to anchor yourself and restore balance to your energy system.

3.   Cultivate Energetic Boundaries: As you explore levitation, it is important to establish energetic boundaries. Protecting your energy field from external influences or draining interactions is crucial for maintaining vitality. Visualize an energetic shield or aura around you, and set the intention to only allow positive and supportive energies to enter your space.

4.   Practice Energy Clearing and Balancing: Regular energy clearing and balancing practices are essential for managing energy drain. Explore techniques such as smudging, energy healing, or chakra balancing to release stagnant or negative energy and restore equilibrium. These practices promote a healthy energetic flow and enhance your ability to sustain levitation states.

5.   Explore Breathwork: Conscious breathwork is a powerful tool for managing energy and enhancing your levitation practice. Deep, intentional breathing oxygenates the body, supports energy circulation, and calms the mind. Incorporate breathwork techniques such as pranayama or rhythmic breathing into your practice to revitalize your energy and enhance your overall levitation experience.

6.   Harness the Power of Intention: Intention is a potent force that can help manage energy drain. Set clear and focused intentions before each levitation session. Align your intentions with the highest good and the energy you wish to cultivate. This conscious direction of energy enhances your ability to harness and sustain levitated states while minimizing energy drain.

7.   Listen to Your Body: Your body serves as a valuable guide in managing energy drain. Pay attention to subtle cues and signals it provides during your levitation practice. If you feel excessive fatigue, strain, or discomfort, it may be a sign that you need to rest or modify your approach. Honor and respect your body's limitations to avoid excessive energy depletion.

8. Seek Energetic Support: When dealing with significant energy drain, seeking energetic support can be beneficial. Connect with experienced energy healers, practitioners, or mentors who can provide guidance, insights, and techniques to help you manage and replenish your energy reserves. Their expertise can assist you in finding a balanced and sustainable approach to levitation.

9. Find a Harmonious Balance: Achieving a harmonious balance between levitation practice and other aspects of your life is vital for managing energy drain. Strive for a healthy integration of your levitation journey into your daily routine. Avoid excessive or obsessive practice that may lead to burnout. Instead, find a rhythm that allows for rest, personal responsibilities, and other activities that bring you joy and fulfillment.

10. Trust the Natural Ebb and Flow: Energy levels naturally fluctuate, and there will be times when energy drain is more pronounced. Trust the natural ebb and flow of your energy and honor your body's need for rest and rejuvenation. Embrace these phases as opportunities for reflection, integration, and self-care. Trust that the energy will replenish, and you will return to your levitation practice with renewed vitality.

By implementing these strategies and maintaining a mindful approach to managing energy drain, you can optimize your levitation practice for long-term sustainability. Remember to prioritize self-care, cultivate energetic boundaries, and listen to your body's wisdom. With a balanced energy system, you can navigate the levitation journey with grace, vitality, and a deep connection to the flow of energy within and around you.

Avoiding Overexertion and Injury for Safe Practice

Levitation is an awe-inspiring practice that requires physical and mental effort. As you explore the realms of defying gravity, it is crucial to prioritize safety and avoid overexertion or injury. In this article, we will discuss essential tips for maintaining a safe and sustainable levitation practice.

1.   Listen to Your Body: Your body is your most valuable guide when it comes to avoiding overexertion and injury. Pay attention to any signs of discomfort, pain, or fatigue during your levitation practice. If something doesn't feel right, take a step back, rest, and reassess. Pushing through pain or ignoring warning signals can lead to injuries or strain.

2.   Start Slow and Progress Gradually: Levitation is a skill that requires gradual progression. Start with basic techniques and gradually increase the intensity or duration of your practice. Allow your body and mind to adapt to the demands of levitation over time. Rushing into advanced techniques without a solid foundation can increase the risk of overexertion and injury.

3.   Warm Up and Stretch: Just like any physical activity, warming up and stretching is essential before engaging in levitation. Warm up your body with light cardio exercises or dynamic movements to increase blood flow, loosen up your muscles, and prepare your body for the practice ahead. Incorporate stretching exercises that target the muscles involved in levitation, such as the core, legs, and back.

4.   Focus on Proper Technique: Maintaining proper technique is crucial to prevent overexertion and reduce the risk of injury. Pay attention to your form, alignment, and body mechanics during each levitation technique. Seek guidance from experienced practitioners or instructors who can help you refine your technique and avoid unnecessary strain on your body.

5.   Incorporate Strength and Conditioning: Building strength and conditioning is essential for a safe and sustainable levitation practice. Focus on strengthening the core muscles, which provide stability and support during levitation. Incorporate exercises such as planks, squats, and back extensions into your fitness routine to enhance the strength and stability of your body.

6.   Take Regular Breaks: Levitation can be physically and mentally demanding. It is important to take regular breaks during your practice sessions. Allow your body time to rest and recover between attempts or techniques. Overexertion can lead to fatigue, decreased concentration, and increased risk of injury. Schedule breaks into your practice routine to maintain a balanced and safe approach.

7.   Stay Hydrated: Proper hydration is essential for overall health and well-being during your levitation practice. Dehydration can lead to muscle cramps, reduced energy levels, and impaired performance. Drink water before, during, and after your practice sessions to stay hydrated and maintain optimal physical and mental function.

8.   Seek Professional Guidance: If you are new to levitation or are unsure about proper techniques and safety measures, consider seeking professional guidance. Work with experienced instructors, trainers, or mentors who can provide guidance on safe and effective levitation practices. They can offer personalized instruction, address any concerns, and help you progress in a safe and controlled manner.

9.   Rest and Recovery: Rest and recovery are essential components of any physical practice, including levitation. Allow your body sufficient time to rest and recover between practice sessions. Incorporate rest days into your schedule to avoid overtraining and give your body the opportunity to repair and rebuild. Prioritizing rest and recovery will help prevent overexertion, reduce the risk of injuries, and promote long-term progress.

10.   Trust Your Intuition: Above all, trust your intuition when it comes to your levitation practice. If something doesn't feel right or if you sense that you are pushing yourself too hard, honor those feelings and adjust accordingly. Your intuition serves as a powerful guide in maintaining a safe and sustainable practice.

By implementing these strategies and prioritizing safety, you can enjoy a fulfilling and injury-free levitation practice. Remember to listen to your body, progress gradually, focus on proper technique, and take care of your physical and mental well-being. With a mindful approach, you can explore the wonders of levitation while minimizing the risk of overexertion and injury.

Levitation and Spiritual Growth: Exploring the Transcendent Connection

Levitation, the act of defying gravity and floating above the ground, has captivated human imagination for centuries. Beyond its physical spectacle, levitation holds profound spiritual significance and potential for personal growth. In this article, we will delve into the intersection of levitation and spiritual growth, exploring how this extraordinary practice can deepen our spiritual connection and catalyze profound inner transformation.

1. Expansion of Consciousness: Levitation opens a gateway to expanded consciousness, transcending the limitations of the physical realm. As we float effortlessly in the air, our perception expands, allowing us to glimpse the interconnectedness of all things. This heightened state of awareness facilitates spiritual growth by awakening us to the infinite possibilities and interconnected nature of the universe.

2. Dissolution of Ego: Levitation challenges our sense of self and ego. When we rise above the ground, we detach from our physical identity and experience a sense of liberation from the constraints of the ego. This detachment allows us to glimpse the transient nature of our ego-driven existence and encourages us to cultivate humility, surrender, and a deeper connection with the divine.

3. Unity with the Divine: Levitation serves as a powerful tool for experiencing unity with the divine. As we elevate ourselves above the earthly plane, we align with higher frequencies and tap into the cosmic energy that permeates all things. This connection with the divine allows us to experience a profound sense of oneness and deepens our understanding of our spiritual nature.

4. Awakening of Intuition: Levitation activates and enhances our intuitive abilities. As we navigate the subtle energies and forces that govern levitation, we develop a heightened sensitivity to the energetic realms. This heightened intuition guides us in our spiritual journey, enabling us to make choices aligned with our higher purpose and navigate the complexities of life with greater clarity and wisdom.

5. Integration of Mind, Body, and Spirit: Levitation requires the harmonious integration of mind, body, and spirit. It invites us to cultivate a deep awareness of our physical, mental, and emotional states, fostering a holistic approach to self-care and well-being. This integration paves the way for spiritual growth by aligning our thoughts, actions, and intentions with our spiritual aspirations.

6. Inner Transformation and Healing: Levitation can catalyze profound inner transformation and healing. As we engage in this transcendent practice, we may encounter unresolved emotions, limiting beliefs, and energetic blockages. By confronting and releasing these obstacles, we create space for profound healing and personal growth. Levitation becomes a transformative journey of self-discovery and self-transcendence.

7. Cultivation of Presence: Levitation demands a deep presence and focused attention. In the state of weightlessness, distractions fade away, and we are fully immersed in the present moment. This cultivation of presence is a powerful spiritual practice that enhances our ability to connect with the divine, experience inner peace, and find meaning in every aspect of life.

8. Service and Compassion: Levitation can inspire a deep sense of service and compassion. As we expand our consciousness and experience the interconnectedness of all beings, we become more attuned to the suffering of others. This heightened awareness fuels our desire to contribute to the well-being of humanity and the planet. Levitation becomes a catalyst for acts of kindness, service, and the cultivation of empathy.

9. Integration into Daily Life: The spiritual growth derived from levitation is not confined to the moments of floating above the ground. The insights, expanded consciousness, and heightened awareness we experience during levitation can be integrated into our daily lives. We carry the lessons learned and the deepened connection with the divine into our relationships, work, and every aspect of our existence.

10. Transcendence of Boundaries: Ultimately, levitation offers us a glimpse of our limitless potential as spiritual beings having a human experience. It invites us to transcend the limitations of the physical world, expand our consciousness, and embrace our divine nature. Levitation becomes a transformative path, guiding us towards spiritual growth, self-realization, and a deeper understanding of the interconnectedness of all life.

As we embark on the journey of levitation, let us embrace its spiritual dimensions. Let us recognize levitation as a sacred practice that invites us to transcend the ordinary and discover the extraordinary within ourselves. Through levitation, we embark on a path of spiritual growth, self-discovery, and profound transformation.

Levitation as a Spiritual Practice: Connecting with the Divine

Levitation, the act of defying gravity and floating above the ground, holds a profound spiritual significance that extends beyond its physical demonstration. For centuries, seekers, mystics, and spiritual practitioners have recognized levitation as a powerful tool for deepening their connection with the divine. In this article, we will explore how levitation can be embraced as a spiritual practice, facilitating spiritual growth, inner transformation, and an expanded consciousness.

1.   Alignment with Higher Frequencies: Levitation enables us to align with higher frequencies and tap into the subtle realms of energy and consciousness. As we transcend the limitations of the physical body and rise above the earthly plane, we enter a state where the veils between dimensions thin. This alignment with higher frequencies facilitates a deeper connection with the divine and opens the door to profound spiritual experiences.

2.   Surrender and Trust: Levitation calls for surrender and trust in the unseen forces that govern our existence. By letting go of our attachment to the physical realm and surrendering to the flow of divine energy, we cultivate a sense of trust in the universe and the wisdom of the divine plan. This surrender allows us to release control and embrace the guidance and support of higher powers.

3.   Stillness and Presence: Levitation requires a state of stillness and presence. As we float above the ground, our focus shifts from the external world to the inner realms of consciousness. This state of stillness and presence allows us to quiet the mind, transcend distractions, and cultivate a deep sense of awareness. In this state, we can access profound insights, intuitive guidance, and a profound connection with the divine.

4.   Transcendence of Ego: Levitation invites us to transcend the limitations of the ego and the illusion of separation. As we rise above the ground, we detach from our physical identity and the ego-driven narratives that define us. This detachment fosters a sense of unity and oneness with all beings, reminding us of our inherent interconnectedness and the divine essence that resides within each of us.

5.   Expansion of Consciousness: Engaging in levitation expands our consciousness and allows us to glimpse the vastness of existence. In this elevated state, our perception shifts, and we become aware of the multidimensional nature of reality. This expanded consciousness brings a deeper understanding of universal truths, the interconnectedness of all life, and the limitless potential of our spiritual being.

6.   Inner Transformation and Healing: Levitation can catalyze profound inner transformation and healing. As we engage in this spiritual practice, we may encounter and release deep-seated fears, limiting beliefs, and emotional blockages. This process of inner purification allows us to align with our true essence, heal past wounds, and embrace our highest potential.

7.   Connection with Spiritual Guides: Levitation can create a bridge between the physical realm and the realm of spiritual guides and higher beings. In this elevated state, we may experience direct communication or guidance from spiritual guides, ascended masters, or angels. This connection provides wisdom, support, and spiritual insights that can guide our journey of self-discovery and spiritual evolution.

8.   Integration into Daily Life: The spiritual practice of levitation extends beyond the moments of floating above the ground. The insights, expanded consciousness, and connection with the divine that we experience during levitation can be integrated into our daily lives. We carry the profound lessons and transformative experiences into our relationships, work, and interactions, embodying the principles of love, compassion, and higher consciousness.

9.   Unity of Body, Mind, and Spirit: Levitation demands the integration and harmony of body, mind, and spirit. To achieve levitation, we must cultivate physical strength, mental focus, and spiritual alignment. This unity of body, mind, and spirit not only supports our levitation practice but also fosters overall well-being and a deep sense of wholeness.

10. Service and Contribution: Levitation, as a spiritual practice, inspires us to embody service and contribute to the well-being of humanity and the planet. As we deepen our connection with the divine and awaken to the interconnectedness of all life, we feel compelled to make a positive impact. Levitation becomes a catalyst for acts of kindness, compassion, and the realization of our role as conscious co-creators in the world.

In conclusion, levitation is not merely a physical phenomenon but a profound spiritual practice that can ignite our inner spark, deepen our connection with the divine, and facilitate spiritual growth. Through alignment with higher frequencies, surrender, presence, and inner transformation, levitation becomes a sacred journey of self-discovery, expanded consciousness, and unity with the divine. Embracing levitation as a spiritual practice opens the doors to extraordinary experiences, profound insights, and a deepening of our spiritual path.

Connecting with Higher Realms

Levitation, the ability to defy gravity and float above the ground, has long been associated with connecting with higher realms of existence. Throughout history, individuals who have experienced levitation have often reported profound spiritual experiences and encounters with higher beings. In this article, we will explore the concept of connecting with higher realms through levitation and delve into the transformative potential of such experiences.

1.   Opening the Gateway: Levitation serves as a powerful gateway to higher realms of consciousness and spiritual dimensions. As we rise above the earthly plane, we transcend the limitations of the physical world and enter a realm where the veils between dimensions thin. This heightened state of being creates an opportunity for direct connection and communication with higher beings, spirit guides, ascended masters, and other entities that exist in higher realms.

2.   Expanded Awareness: Levitation expands our awareness beyond the confines of our physical senses. As we float in a weightless state, our perception expands, enabling us to tap into subtle energies and vibrations that are often imperceptible in our ordinary state of being. This expanded awareness allows us to tune into the frequencies and realms where higher beings reside, facilitating communication, guidance, and profound spiritual experiences.

3.   Vibrational Alignment: Levitation requires a deep vibrational alignment with higher frequencies. In order to rise above the ground, we must attune ourselves to the energetic vibrations that govern levitation. This process of aligning our own energy with the higher frequencies creates a resonance that allows us to bridge the gap between the physical and spiritual realms, facilitating a connection with higher beings and realms of consciousness.

4.   Direct Communication: Many individuals who have experienced levitation report direct communication with higher beings during their floating state. This communication can take various forms, including telepathic messages, intuitive insights, visions, or even auditory experiences. These encounters often impart wisdom, guidance, and profound spiritual teachings that can accelerate our personal growth and spiritual evolution.

5.   Spiritual Guidance: Levitation can serve as a powerful tool for receiving spiritual guidance. When we elevate ourselves above the ground, we create a receptive space within us, allowing us to receive insights, inspiration, and guidance from higher realms. This guidance can provide clarity on our life path, offer solutions to challenges, and illuminate our spiritual journey, leading us towards greater self-realization and alignment with our soul's purpose.

6.   Healing and Transformation: Connecting with higher realms through levitation can facilitate profound healing and transformation. The higher beings and energies encountered in these realms often carry a profound healing presence that can assist in releasing energetic blockages, clearing past traumas, and catalyzing deep inner transformation. This healing process can lead to a greater sense of wholeness, well-being, and alignment with our true self.

7.   Expansion of Consciousness: Levitation invites us to expand our consciousness and transcend the limitations of our physical existence. By experiencing the phenomenon of defying gravity, we challenge our preconceived notions of reality and open ourselves to the vastness of the universe. This expanded consciousness allows us to perceive the interconnectedness of all things and gain insights into the nature of existence, the purpose of life, and the interconnected web of consciousness that we are all part of.

8.   Integration into Daily Life: The experiences and insights gained from connecting with higher realms through levitation are not meant to be confined to the floating state alone. The teachings, guidance, and expanded awareness can be integrated into our daily lives, influencing our thoughts, actions, and interactions with others. We can carry the wisdom gained from these experiences into our relationships, work, and spiritual practices, fostering a deeper connection with higher realms in all aspects of our lives.

In conclusion, levitation can serve as a powerful catalyst for connecting with higher realms of existence. Through the expansion of awareness, vibrational alignment, direct communication, and spiritual guidance, levitation offers us a profound opportunity to transcend the limitations of the physical world and tap into the infinite realms of consciousness. By embracing the transformative potential of these experiences and integrating them into our daily lives, we can deepen our spiritual connection, gain profound insights, and embark on a journey of personal and spiritual growth.

Exploring Consciousness and Expanded Awareness

Levitation, the ability to defy gravity and float above the ground, opens up a doorway to explore consciousness and experience expanded awareness. This phenomenon has captured the imagination of spiritual seekers, mystics, and scientists alike, as it offers a unique perspective on the nature of reality and the potential of the human mind. In this article, we will delve into the fascinating realm of consciousness exploration and expanded awareness through the practice of levitation.

1.   Transcending Physical Boundaries: Levitation challenges our conventional understanding of the limitations of the physical body and the laws of physics. By floating above the ground, we transcend the boundaries of gravity and the constraints of the material world. This extraordinary experience prompts us to question the nature of reality and invites us to explore the realms beyond what is visible to our physical senses.

2.   Heightened States of Consciousness: Engaging in levitation can induce altered states of consciousness that are conducive to expanded awareness. As we detach from the pull of gravity and enter into a weightless state, our consciousness shifts, allowing us to access different levels of awareness. These altered states can range from deep relaxation and meditative states to heightened clarity, intuitive insights, and even mystical experiences.

3.   Tapping into Subtle Energies: Levitation involves working with subtle energies that are beyond the scope of our ordinary perception. By tuning into these energies and learning to manipulate them, we gain access to a deeper understanding of the energetic fabric of the universe. This exploration of subtle energies opens the doors to expanded awareness, enabling us to perceive the interconnectedness of all things and the underlying energetic patterns that shape our reality.

4.   Direct Experience of Unity: Levitation offers a direct experience of unity and interconnectedness. As we rise above the ground, we transcend the illusion of separation and experience a profound sense of oneness with the universe. This experience can be deeply transformative, as it dissolves the boundaries between self and other, revealing the inherent interconnectedness of all beings and the underlying unity that permeates everything.

5.   Accessing Higher Dimensions: Levitation provides a unique opportunity to access higher dimensions of consciousness. As we float above the ground, we can tap into realms of existence beyond the physical plane. These higher dimensions may include spiritual realms, astral planes, or other dimensions of consciousness where expanded knowledge, wisdom, and spiritual insights reside. By exploring these dimensions, we expand our awareness and gain access to profound truths and experiences.

6. Amplifying Intuition and Psychic Abilities: The practice of levitation can amplify our intuitive and psychic abilities. By entering into altered states of consciousness and expanding our awareness, we become more attuned to subtle information and energetic frequencies. This heightened sensitivity allows us to tap into our intuitive faculties, receive guidance, and access information that may be beyond our rational understanding. Levitation can serve as a catalyst for the development and refinement of these intuitive and psychic abilities.

7. Enhancing Spiritual Growth: Levitation can be a powerful tool for spiritual growth and self-realization. The exploration of consciousness and expanded awareness opens up new horizons of understanding, enabling us to transcend limited beliefs, expand our perception of reality, and deepen our connection with the divine. Through levitation, we can embark on a profound journey of self-discovery, spiritual transformation, and the realization of our highest potential.

8. Integration into Everyday Life: The insights gained from exploring consciousness and expanded awareness through levitation can be integrated into our everyday lives. The expanded perspectives, heightened intuition, and deeper understanding of the nature of reality can influence our thoughts, actions, and interactions with others. We can bring the wisdom and insights gained from levitation into our relationships, work, and spiritual practices, fostering a more conscious and awakened way of living.

In conclusion, levitation serves as a gateway to explore consciousness and experience expanded awareness. By transcending physical boundaries, accessing higher dimensions, and amplifying intuitive abilities, we embark on a journey of self-discovery and spiritual growth. Levitation invites us to question the nature of reality, tap into the interconnectedness of all things, and embrace a more expanded and conscious way of being.

Levitation in Everyday Life

Levitation, the ability to float above the ground and defy gravity, may seem like a phenomenon reserved for extraordinary individuals or mystical experiences. However, the principles and practices of levitation can be integrated into our everyday lives, bringing a sense of lightness, freedom, and expanded consciousness to our daily experiences. In this article, we will explore how levitation can be incorporated into our everyday lives, enhancing our well-being, mindset, and overall sense of being.

1.   Cultivating Lightness of Being: Levitation reminds us to cultivate a sense of lightness in our daily lives. Just as we rise above the ground during levitation, we can elevate our mindset and let go of heaviness, both physical and emotional. By releasing unnecessary burdens, worries, and attachments, we create space for joy, spontaneity, and a greater sense of freedom in our everyday experiences.

2.   Embracing Playfulness and Creativity: Levitation invites us to tap into our inner child and embrace a spirit of playfulness. We can infuse our daily routines and tasks with a sense of light-heartedness and creativity. Whether it's finding new ways to approach challenges, incorporating fun and imaginative elements into our work, or simply taking breaks to engage in activities that bring us joy, levitation reminds us to infuse our lives with a sense of play and exploration.

3.   Mindful Presence and Gratitude: Levitation requires focused attention and presence in the present moment. We can apply this principle to our everyday lives by practicing mindfulness and gratitude. Being fully present in each moment allows us to appreciate the beauty and richness of life, while gratitude shifts our perspective towards positivity and abundance. By cultivating mindful presence and gratitude, we can enhance our overall well-being and experience a sense of levity in even the simplest of moments.

4.   Cultivating Inner Balance: Levitation teaches us the importance of inner balance and alignment. In our daily lives, we can seek harmony between our physical, mental, and emotional aspects. This involves nurturing our physical health through exercise, balanced nutrition, and self-care, while also attending to our mental and emotional well-being through practices like meditation, self-reflection, and stress management. When we cultivate inner balance, we create a solid foundation for navigating life's challenges with grace and ease.

5.   Embracing Positive Energy: Levitation is intimately connected with the flow of energy. In our everyday lives, we can consciously choose to surround ourselves with positive energy and create environments that support our well-being. This can involve spending time in nature, engaging in activities that bring us joy, connecting with uplifting and supportive individuals, and cultivating a positive mindset. By consciously choosing positive energy, we invite a sense of lightness and vibrancy into our lives.

6.   Expanding Consciousness: While levitation may not be a daily occurrence for most of us, we can still cultivate expanded consciousness in our everyday lives. This can be done through practices such as meditation, breathwork, mindfulness, and engaging in activities that promote personal growth and self-discovery. By expanding our awareness and exploring the depths of our consciousness, we open ourselves to new perspectives, insights, and a deeper understanding of ourselves and the world around us.

7.   Inspiring Others: Our embodiment of levitation principles in everyday life can inspire and uplift others. By radiating positivity, lightness, and a sense of expanded consciousness, we become beacons of inspiration for those around us. Our presence and way of being can have a ripple effect, encouraging others to embrace a more conscious and joy-filled way of living.

In conclusion, levitation is not confined to extraordinary or rare moments. By incorporating the principles of levitation into our everyday lives, we can cultivate lightness, playfulness, mindfulness, and expanded consciousness. Levitation becomes a metaphor for living with grace, freedom, and a sense of awe, transforming our daily experiences and infusing them with joy and meaning.

Practical Applications of Levitation

Levitation, the ability to defy gravity and float above the ground, may seem like a mystical phenomenon, but it also has practical applications that can benefit various aspects of our lives. Beyond its spiritual and metaphysical implications, levitation techniques and principles can be applied in practical ways to enhance our well-being, improve physical performance, and even contribute to scientific advancements. In this article, we will explore some of the practical applications of levitation and how they can positively impact our lives.

1.   Physical Therapy and Rehabilitation: Levitation techniques can be utilized in physical therapy and rehabilitation settings to assist individuals in regaining strength, balance, and mobility. By reducing the weight-bearing load on injured or weakened limbs, levitation can provide a safe and controlled environment for exercises and movements, allowing for more effective rehabilitation and faster recovery.

2.   Sports Performance Enhancement: Levitation techniques can be beneficial for athletes and individuals involved in sports activities. By developing strength, control, and balance through levitation exercises, athletes can improve their performance in various sports, such as gymnastics, martial arts, dance, and acrobatics. Levitation can enhance body awareness, agility, and spatial orientation, leading to improved athletic skills and abilities.

3. Meditation and Stress Relief: Levitation techniques can be incorporated into meditation practices, promoting relaxation, stress relief, and mental well-being. By experiencing the sensation of floating and weightlessness, individuals can achieve deep states of relaxation and enter into a meditative state more easily. Levitation can serve as a tool for calming the mind, reducing stress, and fostering a sense of inner peace and tranquility.

4. Innovative Transportation Systems: Levitation principles have been explored in the field of transportation, leading to the development of innovative transportation systems such as magnetic levitation (maglev) trains. Maglev trains use magnetic forces to levitate above the tracks, eliminating friction and allowing for faster and more energy-efficient transportation. This technology has the potential to revolutionize the way we travel and can contribute to sustainable and high-speed transportation solutions.

5. Scientific Research and Experiments: Levitation techniques are also employed in scientific research and experiments, particularly in the fields of physics, materials science, and chemistry. Levitation can be used to suspend objects, including liquids and solid materials, in mid-air, providing a controlled environment for studying their properties and behavior. This allows scientists to observe and analyze materials without the influence of gravity, leading to new discoveries and advancements in various scientific disciplines.

6.   Entertainment and Performance Arts: Levitation has captivated audiences for centuries and continues to be a popular element in entertainment and performance arts. Magicians and illusionists incorporate levitation tricks and illusions into their acts, creating awe-inspiring experiences for spectators. Levitation performances can inspire wonder, stimulate the imagination, and create a sense of magic and astonishment.

7.   Creative and Artistic Expression: Levitation techniques can be utilized as a form of creative and artistic expression. Artists, photographers, and filmmakers often incorporate levitation concepts and imagery into their work to convey a sense of freedom, transcendence, and dreamlike aesthetics. Levitation can be a powerful metaphor for breaking free from limitations, defying conventions, and exploring the boundless possibilities of human imagination.

In conclusion, while levitation is often associated with mystical and spiritual realms, its practical applications extend beyond the metaphysical. From physical therapy and sports performance enhancement to transportation systems and scientific research, levitation techniques and principles have the potential to improve various aspects of our lives and contribute to technological advancements. Embracing the practical applications of levitation can unlock new possibilities, enhance well-being, and inspire innovation in diverse fields.

Levitation for Healing and Well-being

Levitation, the ability to rise above the ground and defy gravity, has long been associated with mystical and supernatural powers. However, beyond its mystical allure, levitation techniques and principles can be harnessed for healing and enhancing overall well-being. In this article, we will explore the concept of levitation as a tool for healing and well-being, and how it can positively impact our physical, emotional, and spiritual health.

1.   Energetic Balance and Alignment: Levitation techniques can assist in achieving energetic balance and alignment within the body. By practicing levitation exercises, individuals can become more attuned to their energy centers, often referred to as chakras, and work towards clearing any blockages or imbalances. This can promote the free flow of energy throughout the body, supporting overall health and vitality.

2.   Stress Reduction and Relaxation: Levitation techniques can serve as a powerful tool for stress reduction and relaxation. The act of floating and experiencing weightlessness can induce a state of deep relaxation, allowing the body and mind to release tension and stress. Levitation can provide a respite from the demands of everyday life, promoting a sense of calm and inner peace.

3.   Emotional Healing and Release: Levitation can facilitate emotional healing and release by creating a safe space for individuals to explore and process their emotions. Floating above the ground can evoke a sense of lightness and freedom, enabling individuals to let go of emotional burdens and find emotional release. This can contribute to emotional healing, self-discovery, and a greater sense of emotional well-being.

4.   Enhanced Mind-Body Connection: Levitation techniques encourage a heightened mind-body connection. As individuals practice levitation, they become more aware of the subtle movements and sensations within their bodies. This increased awareness allows for a deeper understanding of the body's needs and promotes a sense of self-care. By nurturing the mind-body connection, individuals can cultivate a greater sense of well-being and harmony.

5.   Accessing Higher States of Consciousness: Levitation can serve as a gateway to accessing higher states of consciousness. When individuals float above the ground, they often enter into a meditative state, allowing for expanded awareness and connection to higher realms of consciousness. This can foster spiritual growth, self-transformation, and a sense of interconnectedness with the universe.

6.   Physical Healing and Pain Management: Levitation techniques can be used to support physical healing and pain management. By reducing the pressure on the body and joints, levitation can alleviate physical strain and promote relaxation. This can be particularly beneficial for individuals dealing with chronic pain, injuries, or physical limitations. Levitation can complement other healing modalities and contribute to overall physical well-being.

7. Spiritual Growth and Awakening: Levitation can be a catalyst for spiritual growth and awakening. Through the practice of levitation, individuals can expand their consciousness, tap into their inner wisdom, and explore the depths of their spirituality. Levitation can open doors to profound spiritual experiences, leading to personal transformation, expanded awareness, and a deeper connection to the divine.

In conclusion, levitation can be harnessed as a powerful tool for healing and well-being. From promoting energetic balance and relaxation to facilitating emotional healing and spiritual growth, levitation techniques offer a holistic approach to enhancing our overall health and vitality. By embracing levitation as a healing practice, individuals can embark on a transformative journey towards greater well-being, self-discovery, and connection to the world around them.

Levitation for Creative Expression

Levitation, the ability to float above the ground and defy gravity, not only captures our imagination but also serves as a source of inspiration for creative expression. The concept of levitation can be harnessed as a powerful tool for artists, writers, musicians, and other creative individuals to explore new realms of imagination and bring their visions to life. In this article, we will delve into the idea of levitation as a catalyst for creative expression and how it can unlock boundless possibilities for artistic endeavors.

1.   Symbolism and Metaphor: Levitation carries rich symbolism and metaphorical meanings that can be incorporated into various art forms. Artists can use levitation as a visual representation of freedom, transcendence, and breaking free from limitations. The image of floating above the ground can symbolize liberation from societal constraints, the unleashing of one's true potential, or the ability to rise above adversity. By incorporating levitation symbolism, artists can evoke powerful emotions and convey deeper messages in their work.

2.   Exploring Unconventional Perspectives: Levitation allows artists to explore unconventional perspectives and challenge the boundaries of reality. By depicting figures or objects suspended in mid-air, artists can create compositions that defy gravity and challenge the viewer's perception of the physical world. This unique perspective can create a sense of wonder and intrigue, opening up new possibilities for artistic expression.

3.   Conveying Emotions and States of Mind: Levitation can be used as a visual metaphor to convey a range of emotions and states of mind. The sensation of floating can represent feelings of joy, euphoria, or serenity. Conversely, levitation can also depict a sense of vulnerability, instability, or uncertainty. Through the depiction of levitation, artists can capture and communicate complex emotions and psychological states in a visually compelling way.

4.   Fusion of Realism and Fantasy: Levitation provides an opportunity to merge the realms of realism and fantasy in artistic creations. By incorporating levitating elements into realistic settings or depicting fantastical scenes where gravity is defied, artists can create a captivating blend of the ordinary and the extraordinary. This fusion of reality and imagination adds depth and intrigue to artistic works, stimulating the viewer's imagination and inviting them into a world of enchantment.

5.   Performance and Dance: Levitation techniques can be integrated into performance arts, such as dance, theater, and circus. Dancers can incorporate levitation movements into their routines, defying gravity with graceful and fluid motions. The act of levitation in performance arts adds an element of awe, captivating the audience and creating a sense of wonder and astonishment.

6.   Inspiring Creative Flow: The concept of levitation can serve as a source of inspiration and catalyst for creative flow. Artists can draw upon the idea of weightlessness and freedom to tap into their creative potential and explore new ideas. The sensation of floating can evoke a sense of lightness and expansive thinking, enabling artists to break free from creative blocks and access new realms of imagination.

7. Visualizing Dreams and Aspirations: Levitation can be used as a visual tool to represent dreams, aspirations, and the pursuit of goals. Artists can depict individuals or objects floating towards the sky, symbolizing the act of reaching for dreams and pushing the boundaries of what is possible. Levitation imagery can inspire viewers to embrace their own aspirations and pursue their passions with determination and courage.

In conclusion, levitation serves as a powerful source of inspiration for creative expression. From symbolism and metaphor to exploring unconventional perspectives and visualizing emotions, levitation offers artists a means to transcend the limitations of reality and venture into the realms of imagination. By incorporating the concept of levitation into their work, artists can unlock endless possibilities for creative expression and captivate audiences with their unique visions.

Printed in Great Britain
by Amazon

37853594R00064